SCIENCE FROM SIGHT TO INSIGHT

SCIENCE FROM SIGHT TO INSIGHT

How Scientists Illustrate Meaning

ALAN G. GROSS AND
JOSEPH E. HARMON

THE UNIVERSITY OF CHICAGO PRESS
CHICAGO AND LONDON

ALAN G. GROSS is professor of communication studies at the University of Minnesota–Twin Cities. He is the author or coauthor of several books, including *The Rhetoric of Science* and *Starring the Text: The Place of Rhetoric in Science Studies.*

JOSEPH E. HARMON works as a science writer and editor at Argonne National Laboratory. He is the coauthor with Alan Gross of several books, including *Communicating Science, The Scientific Literature: A Guided Tour*, and *The Craft of Scientific Communication.*

The University of Chicago Press, Chicago 60637
The University of Chicago Press, Ltd., London
© 2014 by The University of Chicago
All rights reserved. Published 2014.
Printed in the United States of America

23 22 21 20 19 18 17 16 15 14   1 2 3 4 5

ISBN-13: 978-0-226-06820-6 (cloth)
ISBN-13: 978-0-226-06848-0 (paper)
ISBN-13: 978-0-226-06834-3 (e-book)
DOI: 10.7208/chicago/9780226068343.001.0001

Library of Congress Cataloging-in-Publication Data

Gross, Alan G.
  Science from sight to insight : how scientists illustrate meaning / Alan G. Gross and Joseph E. Harmon.
     pages. cm.
  Includes bibliographical references and index.
  ISBN 978-0-226-06820-6 (cloth : alk. paper) — ISBN 978-0-226-06848-0 (pbk. : alk. paper) — ISBN 978-0-226-06834-3 (e-book) 1. Visual communication in science. 2. Communication in science. I. Harmon, Joseph E. II. Title.
  Q223.G772 2014
  501'.4—dc23

                              2013005918

The esthetic image is first luminously apprehended as selfbounded and selfcontained upon the immeasurable background of space or time which is not it. You apprehended it as *one* thing. You see it as one whole. You apprehend its wholeness. That is *integritas*. . . .— Then—said Stephen—you pass from point to point, led by the formal line; you apprehend it as balanced part against part within its limits. . . . In other words, the synthesis of immediate perception is followed by the analysis of apprehension. Having first felt that it is *one* thing you feel now that it is a *thing*. You apprehend it as complex, multiple, divisible, separable, made up of its parts, the result of its parts and their sum, harmonious. That is *consonantia*. . . . [*Claritas* is] the artistic discovery and representation of the divine purpose in anything or a force of generalization which would make the esthetic image a universal one, make it outshine its proper conditions.
—James Joyce, *A Portrait of the Artist as a Young Man* (1916)

The fundamental event of the modern age is the conquest of the world as picture.
—Martin Heidegger, "The Era of the World Picture" (1938)

Seeing is . . . an amalgamation of the two—pictures and language.
—N. R. Hanson, *Patterns of Discovery* (1958)

CONTENTS

Verbal-Visual Interaction in Science

It is remarkable but seldom remarked that scientific communication is unique among the learned enterprises. To see what we mean, flip through the pages of a research article in *Science* or *Nature*. Only in the case of the sciences do practitioners communicate routinely not only through words, but also through tables and images. In literary criticism and science studies, to take two typical scholarly instances, words are ordinarily the sole means of expression, a practice that can sometimes lead to results seriously at odds with reality. Much that passes for Shakespearean criticism, for example, treats its subject as a literary artist. But except in his poems Shakespeare was a literary artist only by accident. By intent, he was a playwright, that is, he made plays, or rather scripts for plays, those combinations of words, actions, actors, sets, and costumes that, with a director's help, constitute the "two hour's traffic" on our stages. It is these verbal-visual amalgams that are the only proper objects of Shakespearean dramatic criticism; to these, Shakespeare is only a contributor. To treat him otherwise is, as Gilbert Ryle (1949) would say, to make a category mistake. Analogously, in much that passes for science studies, scholars eschew the careful analysis of images, despite the fact that the texts with which science studies deal are replete with them. In our view, this is another category mistake, one we hope this book will help correct.

The motives for this communicative mix of words, tables, and images are not hard to find. Tables are an integral part of so many scientific texts because running text is always a poor choice when masses of data must be presented. Visual representations also play an important role in so many scientific texts because establishing scientific truth "almost always requires making something visible and analyzable" (Taylor and Blum 1991, 126). Words, tables, images—each semiotic mode is a vital tool for learning how the world works and for communicating that information to others: indeed, scientific meaning is, we would contend, the product of the interaction among these three elements.[1] The history of scientific communication constitutes our evidence for this claim. We know of no other genre of written communication in which the tabular and visual have held so privileged a place over so long a time. Leonardo da Vinci's notebooks, to use one of the most illustrious early examples, are an intricate web of visual and verbal information on just about every manuscript page, as the following image illustrates.

We are not claiming Leonardo's notebooks are typical of the early scientific literature. On the printed page, it was not until the middle of the nineteenth century that the full integration of the textual, tabular, and visual was realized. But we do contend that, at least from the vantage of the twenty-first-century scientific literature, the evolution in that direction from the earliest days of the printed scientific literature now seems inevitable.

Our book extends a vital but generally neglected insight of historian of geology Martin Rudwick (1976)—that the images embedded in scientific texts have epistemic importance and ought not be ignored or subordinated in exegesis. While in his writings on the history of geology Rudwick has consistently employed this insight, and while historians, philosophers, cognitive scientists, and sociologists have written insightfully about visualization in science, no one, to our knowledge, has attempted a general theory of verbal-visual interaction in the communication of science. Our book is intended as a first step in that direction.

A SHORT HISTORY OF SCIENTIFIC VISUALS AND TABLES

From the sixteenth to the nineteenth century, the illustrated scientific book remained the gold standard of scientific communication. The sixteenth century saw the publication of detailed anatomical drawings in Andreas Vesalius's *De humani corporis fabrica*, and astronomical tables and Euclidean diagrams supporting Nicolaus Copernicus's unfolding mathematical argument for a heliocentric universe in *De revolutionibus*; the seventeenth century, intricate drawings of the microscopic world in Robert Hooke's *Micrographia*, and geometric diagrams related to fundamental laws of motion in Isaac Newton's *Principia*; the eighteenth century, schematics mapping the refraction of light rays through a prism in Newton's *Optics*, and drawings of

---

1. And, importantly, equations—though their integration into a comprehensive theory of scientific communication awaits an expertise that exceeds ours.

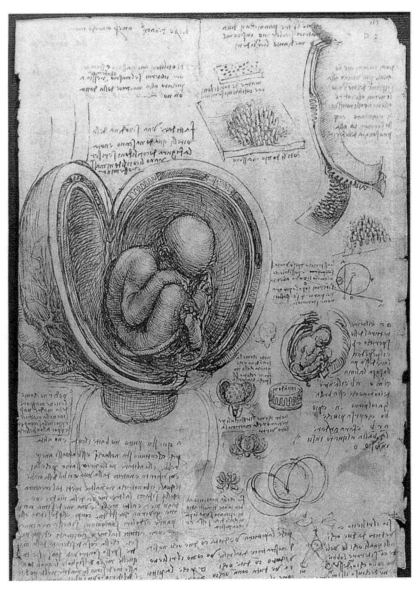

A typical page from Leonardo's notebooks (c. 1510), integrating the verbal and the visual.
Credit: Luc Viatour.

scientific instruments and tables systematically arranging the known universe of chemical elements and compounds in Antoine Lavoisier's *Traité
élémentaire de chimie*; the nineteenth century, geographic strata that reveal
the earth's history in Charles Lyell's *Principles of Geology*, and drawings of
floating microbes produced by fermentation in Louis Pasteur's *Études sur
la bière*.

Visuals and tables have also figured prominently in the main competitor to the scientific book, the research article, which made its debut in 1665
in London and Paris. Until the early nineteenth century, the emphasis in
journal articles was firmly on establishing facts of observation and communicating them by means of verbal descriptions and pictures, primarily
realistic drawings and schematics of natural and man-made objects. Still,
only half the articles from this period contain tables or visuals. Moreover,
the physical integration of figure with text tended to be haphazard in large
part as a result of limitations in printing practices. Figures might appear
gathered together at the end of each article or at the end of the journal. They
might appear in the margins. They might appear within an article on pages
separate from the text or integrated into the page, close to their mention.
They might or might not have descriptive captions. They might or might
not be numbered. Tighter integration of text with the accompanying images, along with titles and numbers, did not occur until well into the nineteenth century when innovations in imaging technology made this practice
economically feasible.

The nineteenth century also added a new visual modality to scientific
books and journal articles: photographs, sometimes taken with the aid of a
telescope or microscope. Prominent nineteenth-century examples include
Edgar Crookshank's photographs of bacteria, Jules Janseen's photographs of
the sun's surface, and Wilhelm Röntgen's X-ray photographs revealing the
skeletal structure of the human hand.[2] Despite their advantages, for technological and economic reasons black-and-white photographs did not become routine in scientific books and articles until the mid-twentieth century. Outside the precincts of popular science journals such as *National
Geographic* and *American Scientist*, the regular appearance of color photographs had to wait for personal computers and the Internet. Now well-
funded journals such as *Nature* and the *Journal of the American Medical
Association*, better known as *JAMA*, are full of them.

The nineteenth century also marks an important conceptual shift in
visualization practices, first noted by Martin Rudwick: geologic schemat-

2. These and other examples of early scientific photographs appear in Keller (2008).

ics shifted from the representation of geologic structures to that of structural causes. With this shift, geologic visuals became "a kind of thought-experiment in which a tract of country is imagined as it would appear if it were sliced vertically along some particular traverse of the topography, and opened along that slice in a kind of cutting or artificial cliff" (Rudwick 1976, 166). By the 1830s the visual language of such sections formed a set of conventions "that were generally accepted and widely understood not only by practicing geologists but also by the wider audience for geology" (172). During this same period other disciplines invented theory-laden means for representing the structures and mechanisms of nature, such as molecular models, evolutionary trees, and electromagnetic field schematics.

During the twentieth century, scientific journals surpassed books as the preferred host for scientific publication. And in these journals, visuals and tables occupy about one-quarter of the average article. Few pages are without them. Of the different visual types on display, Cartesian graphs dominate, a by-product of a major shift in the emphasis of scientific practices from the gathering of observations to the generation and analysis of data, measured or calculated. Cartesian graphs made their first appearance in the scientific literature in the late eighteenth century, invented independently by William Playfair in England and Johann Heinrich Lambert in Germany (Tilling 1975; Wainer 2005). The first visual form with genuine heuristic potential, they contribute to the process of discovery by helping scientists detect changes in data that would not otherwise be apparent. They also contribute to scientific argument by representing law-like relationships as correlations between values on their ordinates and abscissas, and by allowing easy comparisons between theory and experimental data. Although during the eighteenth and nineteenth centuries, this new form for representing data was codified and championed by Playfair and William Whewell in England, James Joseph Sylvester and J. Willard Gibbs in America, and Étienne-Jules Marey in France (Hankins 1999), in our study of the scientific journal literature (Gross et al. 2002), we found no evidence that Cartesian graphs were in general use until the early twentieth century. Until that time, tables remained the principal means for displaying data outside of running text.

The twentieth-century invention of image-producing research equipment—X-ray-based material analyzers, cloud chambers, particle accelerators, electron microscopes, satellite space telescopes, DNA analyzers, and so forth—has moved image creation and analysis closer to the heart of scientific discovery and dissemination. Given the importance of such equipment, the near absence of their depiction in the modern scientific literature may seem odd. Before the twentieth century, after all, such pictures were not uncommon. For example, all of the visuals in Antoine Lavoisier's *Traité élémentaire de chimie* are illustrations of experimental instruments. This is no accident: all the effects Lavoisier writes about in a book that turned chemistry into a modern science are the products of the instruments he uses. But the scarcity of illustrations of instruments in contemporary science is due not to their diminished importance; it is due to their standardization:

> The rapid movement of research equipment from one modification to the next is key to the mode of rapid discovery in which scientists take so much confidence; they feel that discoveries are there to be made along a certain angle of research because the previous generation of equipment has turned up phenomena which are suitable for the intellectual life of the human network. . . . The genealogy of equipment is carried along by a network of scientific intellectuals, who cultivate and cross-breed their technological crops in order to produce empirical results that can be grafted onto an ongoing lineage of intellectual arguments. (Collins 1998, 870–71)

In general, those arguments are fortified by images of the products of research equipment.

Until the late twentieth century, scientific visualization was limited by the printed page, a medium in which color is often too expensive for editorial budgets, space frequently prohibits images large enough to achieve epistemic clarity, and three-dimensional imaging and motion pictures are out of the question. That has changed with the invention of PowerPoint and similar slide projection programs. In the case of these, the subject of our penultimate chapter, the image is absolutely central, an alteration affecting the representation of science. More important on this front, however, is the movement of the scientific literature from the printed page to the computer screen via the Internet, particularly in such preeminent and well-financed journals as *Science, Nature,* and *JAMA*. As will be shown in our last chapter, this new development, combined with the power of the computer for the visual representation of large masses of data, has unleashed a wave of creative image-making unprecedented in the history of scientific visualization.

STATE OF THE FIELD

The classical works in rhetorical criticism paid little heed to the visual element. Neither Aristotle nor Cicero in ancient times, and neither Chaim

Perelman nor Kenneth Burke in modern ones, addressed this issue. Their neglect is understandable. These critics focused on a range of texts in which the visual is of no, or only marginal, importance: orations, sermons, and philosophical and literary works. But despite the obvious communicative importance of the visual in science, in case studies in the rhetoric of science, the emphasis has remained firmly on the analysis and explanation of the verbal. Charles Bazerman's landmark *Shaping Written Knowledge* (1988) reproduces two scientific articles, each of which has a visual, neither of which is discussed; no other scientific images appear. Marcello Pera's *The Discourses of Science* (1994) has only four diagrams—none directly concerned with scientific visualization. Jean Diez Moss's *Novelties in the Heavens* (1993) and Leah Ceccarelli's *Shaping Science with Rhetoric* (2001) contain, respectively, one scientific image, and none. While Greg Myers's *Writing Biology* (1990) and Scott Montgomery's *The Scientific Voice* (1996) contain fourteen and ten scientific images, respectively, these are confined to a single chapter. Jeanne Fahnestock's *Rhetoric Figures in Science* (1999) contains only nine images, while our *Communicating Science* (2002), written with Michael Reidy, contains only sixteen.

In the case of most speeches, most novels, most poems, and most academic articles, exegesis that ignores the visual may be plausibly defended. But in the case of those subjects that depend for their meaning on verbal-visual interaction, science among them, no such defense is plausible. Despite this relative neglect, our review of the academic literature across various disciplines suggests that for the understanding of scientific images we can draw on a considerable body of work from a wide range of disciplines.

We start with a discipline one might think would not have much to offer about the visual—philosophy. Contemporary philosophers of science work within two very different research traditions—the analytic and the phenomenological. From both traditions, we have substantial contributions relevant to scientific visualization. In *Patterns of Discovery* (1958), analytical philosopher Norwood Russell Hanson made three prescient observations: first, that "seeing is a 'theory-laden' undertaking" (19); second, that discovery in science is essentially a matter of *seeing as*, of substituting for the eye we all possess the eye of analysis scientists develop through training and experience; and finally, that the intended meaning of a given visual "is brought out by the verbal context in which it appears" (15). For Martin Heidegger (1938), working in the wake of Edmund Husserl's phenomenology, truth—including scientific truth—is the consequence of "unconcealment," the lifting of a veil. But as scientific truth is a special kind, its visualization is also special: scientific truth is seen as "a calculable nexus of forces," a nexus that reveals itself largely by means of mathematical equations and specialized instruments designed to inquire into nature. It is this process of revelation that turns objects of nature into objects of science. And through this process, the world is "conceived and grasped" as a picture, a picture that is "En-framed," that is, seen through the lens of its mathematicization.

Heidegger's philosophy tends toward the abstract; his books reproduce not a single scientific visual. But some philosophers of science working in the analytical tradition have applied their critical intelligence to episodes in the history of science in which visual communication has figured prominently. Arthur I. Miller (1984) has brought Heidegger's philosophy into the real world by tracing the visual thinking central to the creative breakthroughs of eminent twentieth-century theoretical physicists. Ronald Giere (1996) has given us an instance in the history of geology in which the visual had crucial epistemic significance in settling the controversy over plate tectonics. James Griesemer and William Wimsatt (1989) have demonstrated that germ-plasma diagrams, initially developed by August Weismann in the late nineteenth century, then adopted and modified by others, form a body of evidence to which a theory of evolutionary epistemology can be applied.

In books and articles, cognitive experimentalists and theorists have contributed immeasurably to an increased understanding of how the brain processes verbal and visual stimuli. They have been interested mainly in two topics: the heuristic value of visuals in the thought processes of scientists and the cognitive processes involved in the comprehension of graphs. On the first topic, we have the work of Herbert Simon and such collaborators as Jill Larkin (1987); and on the second, the work of Stephen Kosslyn (1989), Steven Pinker (1983, 1990), and Edward Tufte (2001). William Cleveland and Robert McGill (1984, 1985) may also be counted in this group. In another book, *How Maps Work*, Alan MacEachren (2004) synthesized the published cognitive and semiotic literature into a comprehensive theory of cartographic comprehension.

Some historians have investigated episodes in the history of science that have revolved around the production and analysis of visuals. Julia Voss (2010) has scrutinized how drawings and photographs contributed to Darwin's development of evolutionary theory. David Kaiser (2005) has tracked how Feynman diagrams for representing subatomic particle behavior rapidly infiltrated the practice and teaching of physics after World War II. Peter Galison has addressed the question of "how pictures and counts got to be the bottom-line data of [twentieth-century] physics" (1997, xvii). Martin Rudwick (2005) has traced how both geological verbal and visual commu-

nications radically altered the way eighteenth- and nineteenth-century sa-
vants thought about time and the evolution of the earth.

Several historians have documented the history of scientific visuals
themselves. Michael Friendly and Daniel Denis have produced the most
prominent of these studies, "Milestones in the History of Thematic Cartography, Statistical Graphics, and Data Visualization" (2010). This web-
based database presents an elaborate timeline regarding the history of the
visual representation of quantitative information, starting with the oldest
known map (c. 6200 BCE) and ending with "sparkline" (2004), word-sized
graphs designed to appear in running text.[3] Each milestone on the timeline
has hyperlinks to visuals illustrating it. In a more traditional vein, Thomas
Hankins (1999) has written a history of data graphs, the most prevalent form
of scientific visual communication over the past century. The lavishly illustrated *Maps: Finding Our Place in the World* (Ackerman and Karrow 2007)
offers a historical overview of the oldest form of scientific visualization.
Lorraine Daston and Peter Galison have collaborated on *Objectivity* (2010),
a work that employs scientific atlases to analyze the contingent nature of
their central term.

Sociologists of science have weighed in on the social processes by which
scientists transform into pictures the knowledge generated in the laboratory
or field. In the wake of Michael Lynch's (1985) *Art and Artifact in Laboratory Science*, we have two comprehensive collections of essays: Michael
Lynch and Steve Woolgar's *Representation in Scientific Practice* (1990a) and
Luc Pauwels's *Visual Cultures in Science* (2006). In these works, we find
that sociologists have tape- and video-recorded the shop talk of scientists
creating and interpreting visuals, interviewed them about the subject, and
undertaken case studies into the sociocultural forces out of which scientific
visuals arise. As evident from studies in both the history and sociology of
science, image making in science entails much more than converting numbers or objects into pictures that reveal "truths" about the natural world
untainted by outside influence (Galison 1998; Fleck 1979; Panese 2006).

Finally, there are books emphasizing the aesthetic side of scientific visualization. *Visualizations: The Nature Book of Art and Science* (Kemp
2000) and *Picturing Art, Producing Science* (Jones and Galison 1998) explore the boundary between the knowledge-producing activities of science
and the image-making activities of art. In *Cosmic Imagery: Key Images in*

---

3. March 2010. No doubt Friendly's visual milestones will have long since been updated by
the time our book appears in print.

the History of Science (2008), the full-time physicist and part-time historian of science John Barrow celebrates the image-making tradition in science and mathematics dating back to ancient Greece, Babylonia, and China. Felice Frankel's *Envisioning Science: The Design and Craft of the Science Image* (2002) offers scientists practical advice on creating striking scientific photographs worthy of journal covers. Several fine exhibitions of scientific art have even made the museum circuit, such as *Maria Sibylla Merian & Daughters: Women of Art and Science* (Reitsma 2008) and *Brought to Light: Photography and the Invisible 1840–1900* (Keller 2008). And a book worthy of any coffee table or bookcase is *Portraits of the Mind: Visualizing the Brain from Antiquity to the 21st Century*, by Carl Schoonover (2010). Naturally, all these books emphasize the "pretty pictures" of science. But whether a visual is good, bad, or ugly from a purely artistic perspective is a matter of no particular consequence from a purely scientific perspective. Indeed, the vast majority of contemporary scientific visuals are as plain, some might say relentlessly drab, as the text that accompanies them. Scientific visuals share with much contemporary art a fine disregard for the merely beautiful.

OUR BOOK

These scholarly books and articles offer valuable insights into visual creation, comprehension, communication, and even miscommunication. What they do not offer is a comprehensive theory of just how, by means of the interaction of the verbal and the visual, scientific meaning is communicated. It is no wonder that Peter Galison complained that "we have a large number of empirical case studies but a relatively impoverished understanding of how visualization practices work" (2008, 116). In the present book, we propose to further just such an understanding, grounded in the philosophy of science and synthesized from research in cognitive psychology, semiotics, and argument and narrative theory. We apply the theory we synthesize to scientific books and articles in print, to PowerPoint presentations, and to contemporary scientific articles that exploit the opportunities for creative visual expression afforded by the Internet. In taking a synthetic approach, we hope to overcome a problem in the study of the role of visuals in scientific communication: the general reluctance to pursue a synoptic midlevel theory with the aim of exegetical enlightenment. The highly regarded French theoreticians on the semiotics of scientific illustration, Jean Bertin (1981, 1983) and Françoise Bastide (1990), avoid general formulations. Bertin's semiotic work parallels in purpose the very differently oriented work of Tufte (2001) and Cleveland and McGill (1984): it is theoretical only in the interest of amelioration. Bastide's sensitive analysis assumes but does not make explicit the semiotic theory of visuals driving her work.

Gunther Kress and Theo van Leeuwen have also taken up the challenge of addressing visual communication in *Reading Images* (1996), adopting a theoretical approach that combines semiotics, linguistics, and aesthetics. Their example images cover a broad swath: advertising, movie stills, artistic paintings and drawings, illustrated pages of newspapers and magazines, and even a few technical illustrations. In a follow-up book, *Multimodal Discourse* (2001), they expand their analyzed examples beyond static images to cover sound and motion. While we have found both works inspirational and provocative, we have in the end rejected the approach of the first because its analysis is based too heavily on theories originally designed for verbal exegesis. We have rejected the second for a different reason: we feel that its rapid ascent to high theory is premature.

We begin chapter 1 with Heidegger's hermeneutics, a highly abstract theory in which visualization plays a central role. Despite its focus on visualization, however, Heidegger's philosophy is insufficient as an explanatory framework because it ignores scientific argument, a topic on which analytical philosophy offers us considerable insight. There is another problem with Heidegger's philosophy: it cannot be applied directly to the analysis of scientific visuals. It is a theory of how human beings experience themselves and the world, not a theory of how the mind works in processing scientific visuals. Given this lacuna, we had to find a theory of cognitive psychology compatible with Heidegger's views. Our choice is Allan Paivio's dual coding theory (2007), a formulation that takes seriously the differences between the verbal and the nonverbal, and the constraints and affordances that their perception and cognition entail. In dual coding theory, verbal and nonverbal stimuli are channeled separately and stored apart in the brain; it is their combination that generates meaning. To bring dual coding theory down to the earth of texts, to render it sensitive to both their gross anatomy and histology, we borrow from several sources. To account for the perception of patterns, we have recourse to Gestalt theory. From the literature on scanning and matching, we borrow the means by which readers confronting a table or an image identify potentially semiotic components. From Charles Sanders Peirce (1955), we appropriate the means by which perceptions are interpreted in different contexts, either as *icons* that represent reality, as *symbols* that stand for it, or *indices* that point to causes. For instance, an X-ray image may represent the internal structure of a human hand (icon), stand for all human hands (symbol), or reveal a defect linked to pain (index).

To complete our analysis of the gross anatomy and histology of texts, we add to our formulation argument and narrative theory. According to the synthetic and synoptic theory we propose, readers of scientific tables and visuals perform the following tasks: by means of Gestalt patterning, they perceive structures and their components; by means of regimes of scanning and matching, they identify the components of these structures; by means of semiotics, they interpret them; finally, by argumentative and narrative means, they integrate these meaningful structures and components into semiotic wholes. In the remaining chapters, we hope further to clarify our theory's application to images and tables in scientific texts, to analyze its realization in scientific argument and narrative, and to apply it so as better to understand the changes in scientific visualization over time.

In our second and third chapters, we look at the diverse ways in which scientific tables and visuals have exploited an ecological niche that first emerged in the fifteenth century: the printed page. There is little question that print itself facilitated the growth of science in the Western world, accelerating its progress by orders of magnitude (Eisenstein 1979; Latour 1990). Moreover, the new medium was a spur to graphic creativity. Printing made possible the wide dissemination of visualization by means of woodcuts, engraving, lithography, and much later, photography—technologies that opened the door to creativity in the medium itself. The page became an environment in which a variety of visuals evolved as a consequence of the visual ingenuity scientists consistently displayed.

In chapter 2 we try to bring some sense of order to this magnificent proliferation by proposing a taxonomy of visual communication in science. Our intention is not to alter scientific naming practices, hallowed by tradition. We are content that current terminology—that such terms as *table, figure, graph, diagram,* and *map*—remain in place. But is the periodic table really a table? Is Darwin's one visual representation in the *Origin of Species* a diagram, map, or graph? What do we call a visual representation of a molecular structure? In this chapter, we take a first step to overcome terminological confusion. To do so, we classify tables and figures in terms of the semiotic purpose they serve. The system we employ is based on a single organizing principle, the overriding purpose a representation serves in furthering the arguments scientists make. We classify representations into seven categories: 1) those designed to present and retrieve data, that is, tables; 2) those designed to present data and express data trends, such as line and bar graphs; 3) those designed to express spatial relationships, such as maps and molecular models in chemistry; 4) those designed to express space-time interrelationships, such as Galileo's astronomical representations of the changing positions of the moons of Jupiter; 5) those that use space as a metaphor for a relationship other than spatial, such as flowcharts and circuit diagrams; 6) those whose purpose is virtual witnessing, such as photographs of star clusters or cloud chamber events; and 7) those designed to reveal the function of research equipment and experimental arrangements. For each category we posit, we select what we judge to be a representative example and, using the verbal-visual interaction model presented in chapter 1, show how it communicates science.

Our third chapter traces the evolution of scientific tables and visuals to the point at which Heidegger's philosophy of science applies to them without exception. According to Heidegger, modern science sees the world through a mathematical lens as a "calculable nexus of forces." This status was not achieved simultaneously by all the sciences. In *The Eye of the Lynx* (2002), art historian David Freedberg explores the unsuccessful efforts of Prince Federico Cesi, Galileo's associate, in spearheading the assembly of a visual archive depicting the Old and New World's flora and fauna. Freedberg shows that Cesi's realistic depictions of these in all their variety was, from the point of view of modern science, doomed to failure. Diagrams of the relationships among things were required, not "realistic" depictions. This transformation to Heideggerian science begins with seventeenth-century astronomy, already mathematized in ancient times, though not, as we shall see, precisely in Heidegger's sense. By the end of the first quarter of the seventeenth century, both astronomy and physics had become modern sciences fully Heideggerian in character, studies in which the world is conceived and grasped as a picture of "spatiotemporal magnitudes of motion" (1938, 119). By the middle of the nineteenth century, chemistry and geology also arrived at this point. To align the life sciences with the rest, to make these sciences fully Heideggerian, was to transform the study of living creatures into a search for the calculable nexus of forces that made them what they were. We cannot say with certainty that by the middle of the nineteenth century the life sciences had begun this conceptual journey, but we do contend that the single diagram in Darwin's *Origin of Species* exemplifies a significant step in this direction. At what point the life sciences became mathematical in a strict sense, became actually a matter of numbers, is in any case beside the Heideggerian point; once the world is grasped as a picture in Heidegger's sense, mathematics becomes no more than the actualization of an ever-present possibility. In the final section of this chapter, as an extended example of the Heideggerian transformation of the life sciences, we track the changing nature of the visuals in brain localization research from the late nineteenth through the twentieth century.

In chapter 4, we explore the ways in which words, numbers, and images interact to communicate scientific arguments. We do so within a model that encompasses the contexts of discovery and justification. We contend that the arguments we see in the books and articles that constitute the primary scientific literature are the end points in a process of discovery and justification, a chain of reasoning that leads from hypotheses about the way the world might be to arguments about the way the world is. At each stage of the model of this process that we favor, vigilant problem-solving manifests itself: a rational decision must be made. At stage one, the decision is to attack a particular problem; at stage two, to pursue a particular hypothesis; at stage three, to solicit acceptance of a particular truth claim on the basis of the arguments set out in its favor. From the point of view of the generating scientists, the process is entirely teleological: persuasive argument is always the end point. Many factors influence whether or not fellow scientists are persuaded: elegance, simplicity, predictability, precision, openness to confirmation or disconfirmation, fruitfulness, wide scope, specification of a mechanism, unification of existing facts and theories, fit with existing background beliefs, variety in the sources of supporting evidence, the track record of the authors and their institutional affiliations.

In this chapter, we contend that scientific argument transforms the objects and events it studies. As the discussion about them passes through the stages of the model we advocate, their semiotic status alters: they shift from the iconic and symbolic to the indexical, that is, from representing and generalizing about the world to explaining how the world works. By means of argument, for instance, James Watson and Francis Crick turn X-ray diffraction photographs of DNA into an iconic molecular structure, the double helix. But this shift is also a transformation from the iconic to the symbolic. Because the Watson-Crick model applies to any DNA molecule, each of its realizations stands for the rest. There is a further transformation: the molecule's structure suggests a scheme of replication, a shift in mode from the iconic to the indexical, from description to explanation. Contemporary scientific argument is Heideggerian: it En-frames a picture of the world in which a certain kind of rationality is consistently privileged.

In our analysis of scientific argument as a consequence of verbal-visual interaction, we left undiscussed a family of sciences in which time is the central issue and narrative the final form. While all sciences have as their goal the discovery of processes that repeat themselves over time, only in such "historical" sciences as geology, evolutionary biology, anthropology, and archaeology is the goal also the explanation of events unique in time, those generated by means of such processes. Our concern in chapter 5 is not with ordinary narratives, such as those that present the story behind some scientific discovery, but with "argumentative narratives," structures in which narrative and argument form, as it were, two sides of a Möbius strip. In this chapter, we contrast the work of a geologist who succeeded in telling a credible story and one who did not but whose successors did. Charles Darwin successfully attributed creation of fringing reefs, barrier reefs, and atolls to the same cause: subsidence of large land masses over geological time. By contrast Alfred Wegener unsuccessfully attributed the present position of the continents to their drift over time through an underlying permeable base. It can be argued that Wegener's strident advocacy of this form of continental drift in the final edition of *The Origin of Continents and Oceans* (1929) was no more than outrageous speculation supported by hollow rhetoric; certainly, this was the opinion of many of his geological contemporaries. We argue instead that this distinguished scientist contributed to his science by keeping alive his intuition that the continents did in fact change position over geological time. It is this intuition, we argue, that the advocates of seafloor spreading successfully supported with solid evidence several decades after Wegener's death. In all three cases—Darwin, Wegener, and the seafloor-spreading advocates—the authors tell a story of the earth's geologic history and rely heavily upon spatiotemporal visual representations in making their arguments.

In chapter 6, we explore the ways in which verbal-visual interaction accommodates science to contexts that extend beyond the specialized areas in which it originates. The initial neglect of Mendel's work alerts us to the problem a more encompassing scientific context poses for scientific argument. While the genetics of the common pea in fact mimics animal genetics, it does no justice to the sexual variety that generally characterizes the botanical kingdom; contemporary botanists were right to be skeptical to any generalization of Mendel's new knowledge claim. The continued opposition to Copernicanism alerts us to the problem that a more encompassing social context poses for scientific argument. The Catholic Church was rightly concerned that astronomical might trump theological criteria in the determination of the nature of the universe; only the eventual weakening of the authority of the Church regarding matters of the physical universe would yield anything approaching a widespread acceptance for Copernicanism.

Our example of this dual accommodation concerns the efforts of nineteenth-century geology, evolutionary biology, and anthropology to domesticate the notion of "deep time" for a Victorian audience skeptical of its scientific merit and profoundly disturbed by its social implications. In the latter half of the nineteenth century, geologists such as James Geikie and

evolutionists such as Charles Darwin were faced with a formidable scientific obstacle: Lord Kelvin's calculations of the age of the earth. Kelvin had
argued that according to the second law of thermodynamics, the law of en-
tropy, the earth could not be anywhere near as old as geological or biological
evolution required. Until the discovery of radioactivity, there was no sound
counterargument.

While Kelvin's argument troubled only geologists and evolutionary biologists, anthropologists such as John Lubbock joined their fellow scientists
in facing an equally formidable obstacle. The advocacy of deep time had to
make its way against the tide of a public hostile to the very idea. It was a
hostility deeply rooted in the Judeo-Christian tradition, a set of intellectual
presuppositions and emotional habits according to which human beings are
central, and human time is the only time. How would it be possible to overcome the disenchantment this new view entailed, the danger of alienating
human beings from their own past? To counteract this potentially debilitating effect, geologists, evolutionists, and anthropologists needed to turn the
earth into *our* earth, living things into *our* relatives, and early hominids into
our direct ancestors, to transform those images into *our* images. Images of
geological formations, evolutionary characteristics, and anthropological artifacts are interpretively innocent; the world comes with no labels attached.
When we look at an image of a flint, unless we are experts, we cannot tell
whether its shape is due to human or natural forces, and, if we determine
its maker to be human, we cannot, unless we are experts of a different order,
tell whether it is a tool or a weapon, or whether it was made yesterday or a
million years ago. It is we—or rather the scientists we deputize—who turn
these images into icons of our past, symbols for its vastness, and indices of
past practices. It is these semiotic transformations, which depended on accommodation of a Victorian audience to deep time, that this chapter will
investigate.

The final two chapters of the book concern the ways in which verbal-
nonverbal interactions have evolved in adapting to two newer media: Power-
Point and the Internet. In chapter 7 we deal with PowerPoint, in our view
an important agent of change in the way science is communicated and represented. It has been forcefully argued that this change is for the worse, that
PowerPoint has undermined all oral communication because its numerous
constraints force content into regimented presentations, best exemplified
by bullet-point lists that routinely drive audiences to distraction. But, we
argue, this usually dreary result stems not from PowerPoint but from our
refusal to examine and exploit this medium, to explore its constraints and

affordances fully, sensibly, and responsibly. When its presenters do so, PowerPoint proves to be a medium ideal for the communication of a subject like science, in which verbal-nonverbal interaction is central. We support this hypothesis with an analysis of three PowerPoint presentations aimed at three different audiences with an interest in science: the general public, college teachers and their students, and professional scientists. We show that the crucial unit of analysis is not the individual slide and its images, but the extent to which that slide is integrated into the presentation as a whole. We show, moreover, that the principle by which this integration is achieved changes as the audience does: general audience presentations are best organized by means of narrative, while professional audience presentations are best organized by means of argument, making a new knowledge claim and arranging evidence in support of it. In all cases, audience adaptation is the master variable, determining what counts as the optimal integration of the verbal and the visual into a single message.

In our view, this integration, in which the visual is routinely subordinated to the verbal, has an ontological dimension: the way science is represented is, potentially, altered. We illustrate this alteration by analyzing another PowerPoint aimed at a professional audience, Jonathan Losos's presentation on the adaptive radiation of Caribbean lizards. In this presentation, Losos represents science in a way that deviates from the standard picture: he depicts evolutionary biology as a practice; he shows how that practice transforms the world we experience into a world in which images and data sets stand for natural forces. In Heideggerian terms, Losos's PowerPoint enacts the En-framing of the world, turning it into its mathematization, the consequence of a calculable nexus of forces.

Chapter 8 analyzes the new communicative and epistemic possibilities the Internet creates as a consequence of the studied abandonment of the printed page, and the constraints it imposes on verbal-visual interaction and on the timeliness and character of scientific publication. In exploiting these possibilities, such elite journals as *Science, Nature, Cell, PLoS Biology, Journal of the American Medical Association*, and *New England Journal of Medicine* lead the way. Research approved by scientific communities as provisional knowledge can now be shared with almost no delay. Moreover, the virtual witnessing that Shapin and Schaffer (1985) attribute to Robert Boyle's prolix seventeenth-century prose has become for twenty-first-century scientists a three-dimensional, multisensuous reality in the form of 3-D images that rotate 360 degrees, and graphs that change in real time. Internet articles also include sound, and short films depicting events in the lab or field as well as events simulated by computer modeling. And these images are routinely in color, still a considerable expense in print-only technology.

Not only is the Internet altering the way scientific images picture the world; it is also changing the overall visual presentation of the research article itself. Figures and tables can appear where they belong, not where the exigencies of the printed page require. Subject to the ever-loosening restrictions on copyright by publishers of scientific books and journals, references can be "live"—cited articles can be only a click away. In the case of "newsworthy" articles, general reader video presentations, including interviews with the scientists involved, can be accessed in supplementary materials. In these, we can also see much that had been excluded from public view by the exigencies of print: for example, supplemental tables of data, supplemental methods, and video clips of field and laboratory events. The Internet revolution also has a social dimension: it is now possible for Bacon's *New Atlantis* vision of science as a collaborative enterprise to be more fully realized. Before official "publication," research in the making can be shared for comment with relevant scientific communities. In addition, for the first time since the days of nineteenth-century geology and ornithology, serious amateurs can contribute to science. They can do so by means of Wikipedia-like collaborative multimedia projects with such lofty goals as gathering the acquired knowledge on every species of organism on Earth (Encyclopedia of Life, www.eol.org) and the evolutionary origins of all living things (Tree of Life Web Project, tolweb.org). Finally, journals are taking advantage of visual learning by posting videos of procedures being performed, ranging from the sequencing of DNA to the removal of earwax in a squirming child. On the Internet, what was once static is now dynamic, and what was once black and white is now routinely in color and the product of computerized imaging technology. Visual-verbal interaction has been enhanced beyond print culture's imagining.

END OF THE BEGINNING

After over two decades of thinking and writing about science, we have concluded that our object of study has been seriously distorted by a bias we have shared with our colleagues in philosophy, sociology, and history, a bias in favor of the verbal. This bias has been enforced by years of training in the analysis of verbal texts; for years, it has seemed reasonable to take these as the norm. A bias toward the verbal has been reinforced by a practice of analytical philosophy, the dominant philosophy in the United States and

England. In imitation of theoretical physics, some analytical philosophers have searched sedulously for a "final theory," the ultimate set of axioms from which the world we experience can be deduced. Ambitious programs such as this are bound to find scientific images irrelevant to their enterprise, or rather relevant only to the extent that certain of their characteristics are open to abstraction. But programs of this sort also leave behind the daily practice of most scientists, men and women who are not, after all, analytical philosophers. The daily practice of these scientists includes the scrutiny of such images as false color photographs of the solar corona and the creation of such images as line graphs. We offer no apologies, therefore, in reaching across the philosophical divide to Martin Heidegger, a philosopher who saw clearly that visualization was central to the sciences.

We are certain of two things about our book. First, we are right to have added our voice to those of William Wimsatt, Martin Rudwick, Peter Galison, Michael Lynch, and Martin Kemp, among others, in an effort to spread the news to those in science studies, telling them that they may be marginalizing a characteristic of science that is, in fact, at its center. Second, we feel confident that our theory of verbal-visual interaction, assembled from disparate components of phenomenology, Gestalt psychology, cognitive psychology, scanning and matching theory, and Peirce's semiology, has genuine heuristic potential; it has helped us see what would otherwise have been invisible to us. We also believe that what we have created can be of use to other science studies scholars in understanding scientific visuals better. We hope that our work spurs others in science studies to more nuanced theoretical efforts, and that advances in cognitive psychology and semiology assist them in their endeavor.

CHAPTER ONE

A Framework for Understanding
Verbal-Visual Interaction

[Natural] Philosophy is written in that vast book which stands forever open to our eyes, I mean the universe; but it cannot be read until we have learned the language and become familiar with the characters in which it is written. It is written in mathematical language, and the letters are triangles, circles and other geometrical figures, without which means it is humanly impossible to comprehend a single word.
—Galileo Galilei, *The Assayer* (1623)

Our assumption that scientific communication is a consequence of the interaction between words and images requires us to erect a framework for analysis that takes verbal-visual differences seriously into consideration, and that makes the explanation of their interaction a central concern. Our first task is to provide a philosophical justification for our endeavor. For this we turn to Martin Heidegger, the first philosopher to emphasize the central place of visualization in the sciences. Heidegger's philosophical perspective forms the initial component in our framework. For the second, we rely on the dual coding theory (DCT) of the Canadian cognitive psychologist Allan Paivio. We do not know whether the central premise of DCT, the hardwired dual coding of the verbal and the visual, will survive the test of time, but it is a plausible explanation of the facts currently available, and a plausible translation into the language of cognitive psychology of Heidegger's concern with the epistemic primacy of the visual. The final component in our framework of analysis stems from a limitation we perceive in DCT: it lacks a viable hermeneutic capacity. Its level of abstraction is such that both the fine and the more encompassing structures of scientific communication escape its grasp. In the absence of a well-established hermeneutic theory of verbal-visual interaction on which we can rely, we have called for assistance from the

resources of Gestalt psychology, scanning-and-matching theory, Peircian semiotics, and argument and narrative theory. To indicate its relationship to
DCT, we call our synthesis enhanced dual coding theory (EDCT). According to EDCT, we perceive images as patterns ordered in accord with Gestalt
theory; we identify their components by means of scanning-and-matching
theory; we interpret these and the wholes of which they are a part by means
of Peircian semiotics; finally, we integrate these interpretations into semiotic wholes by means of argumentative and narrative structures.

FIRST COMPONENT: HEIDEGGER'S
PHILOSOPHY OF SCIENCE

It is impossible to exaggerate the influence of Heidegger's phenomenology
in twentieth-century philosophy; one need only mention Hans-Georg Ga-
damer, Maurice Merleau-Ponty, and Emmanuel Levinas. Even in philosophy
of science, still dominated by the analytical tradition, Heidegger has his firm
adherents: Patrick Heelan (1983), in his writings on modern physics and visual spaces, and Don Ihde (1991, 1998), in his writings on the visual hermeneutics of technology, acknowledge their debt on virtually every page. And
the debt may be there, even if unacknowledged. In Bruno Latour's *Science in
Action* (1987), while Heidegger is not in the index, he is in the wings, super-
intending the production, despite Latour's protestations to the contrary. And
how different are Kuhn's paradigms and Heidegger's *Ge-Stell*, or En-framing?

Still, there is a barrier to untrammeled use: scholars wishing to rely
on Heidegger's philosophy cannot ignore the question of whether his unre-
pentant Nazism infects those portions of his work they intend to employ
(Faye 2009). Gottlob Frege may seem like a parallel case. Are his obnoxious
political views to prevent us from acknowledging his fundamental contributions to the philosophy of language? At least one eminent philosopher,
Michael Dummett, did not think so, though his discovery of Frege's racism
shocked his conscience. But Heidegger's case is importantly different, as
there is little question that he saw the Nazi movement as the realization of
central aspects of *Being and Time* (1962a) or that his views of science were
harnessed to a polemic whose target was the technological supremacy of
a modern world founded on capitalism and democratic principles (Wolin
1991; Fleischaker 2008). But to say this is not to say that many of his insights into hermeneutics in *Being and Time* (1962a) and into science in his
later essays are necessarily expressions, respectively, of his Nazism and his
reactionary views. In our view, they are not; Heidegger's Nazism is irrelevant to the case we wish to make.

In another sense, however, the disruptions of the Nazi era are relevant to our argument because they concern the split between phenomenology, with its highly specialized vocabulary and emphasis on the interaction of consciousness with the world, and analytical philosophy, with its distrust of metaphor and emphasis on the derivation of fundamental principles by mathematics and logic. It would be unrealistic to expect phenomenologists and analytical philosophers to agree on fundamentals. But it would not be unreasonable to expect that they would treat each other with respect, that they would converse with the expectation that their conversations might be mutually beneficial. At least one analytical philosopher, Michael Friedman, feels that the 1929 encounter between Heidegger and Ernst Cassirer in Davos, Switzerland, an encounter at which Rudolf Carnap was present, was characterized by such mutual respect. It was a rapprochement that was not to last: only four years later, the rise of Nazism left Heidegger as the only philosopher of distinction in Germany.

In Friedman's view, a contingent circumstance led to what now seems like an unbridgeable divide: "the thoroughgoing intellectual estrangement of these two traditions, their almost total lack of mutual comprehension is a product of the National Socialist seizure of power" (2000, 156). While our ambition does not extend to healing this breach, its contingent origin encourages us to harness to our own purposes any insights into our subject, regardless of their source. We take from Heidegger the epistemic and ontological primacy of the visual to modern science, a primacy in which the analytical tradition has no interest; we take from the analytical tradition the reconstruction of science as a network of propositions, an enterprise in which Heidegger had scant interest. Only in this way do we think that we can do full justice to scientific communication as the argumentative and narrative product of the interaction of the verbal and the visual.

Heidegger's philosophy of science seems to us a good choice when the object of study is scientific visuals: his is a phenomenology centered on sight and insight (Glazebrook 2000). We do not mean that *Being and Time* and subsequent works are dominated by visual metaphors, although that is certainly the case; we mean rather that the significance of Heidegger's philosophy is inseparable from these metaphors. For most philosophers, these are, like Kant's dove, merely illustrative. Like the philosopher tempted by the snares of pure reason, Kant's dove "might imagine [wrongly] that its flight would be still easier in empty space" (1787, 47). For Heidegger, however, metaphor is a philosophical tool, a means of reaching beyond the conceptual prison of denotative language. In Heidegger's case, this procedure has an unfortunate result: it leads to prose that is equally rebarbative in

German and in English translation. We will deal with this problem by providing a commentary that, we hope, illuminates the passages we quote, followed by Heidegger's prose in translation as a check on the validity of our
commentary. We are under no illusions that our exegesis will result in a
consensus on the part of all readers. We claim only that our interpretations
are plausible, and that they support our claims for the relevance of Heidegger's philosophy of science to the task we have set for ourselves.[1]

Meaning, for Heidegger, is conceived in terms of sight. In typical fashion, he weaves literal and metaphorical discourse into a single philosophical
configuration, playing on the German for "seeing" (*sehen*) and for "the visible" (*sichtig*). In order to reach a point where something is explicitly understood, he asserts, we must take apart what is "circumspectively [*umsichtig*]
ready-to-hand," that is, whatever is available for use, until we reach a point
where it has "the structure of something as something." It is at that point
"explicitly understood." As Heidegger elaborates,

> all preparing, putting to rights, repairing, improving, rounding-out, are
> accomplished in the following way: we take apart in its "in-order-to"
> that which is circumspectively [um**sicht**ig] ready-to-hand, and we con-
> cern ourselves with it in accordance with what becomes visible [***sichtig***]
> through this process. That which has been circumspectively [*um**sicht**ig*]
> taken apart with regard to its "in-order-to," and taken apart as such—
> that which is *explicitly* understood—has the structure of *something as
> something.* (1972, 148–49; 1962a, 189; emphasis his)[2]

---

1. Our claim of plausibility is further attenuated by the character of Heidegger's oeuvre.
After the first part of *Being and Time* (1962a, first published in German in 1927), we have from
him not the second part, but only lecture series and occasional essays. In two lecture series—
The Phenomenological Interpretation of Kant's "Critique of Pure Reason" (1927–28) and *What
Is a Thing?* (1935–36)—he reflects on science; he also reflects on science in the following essays:
"The Era of the World Picture" (usually translated as "The Age of the World Picture") (1938),
"What Is Technology?" (usually translated as "The Question Concerning Technology") (1953a),
and "Science and Reflection" (1953b). We interpret these intellectual excursions as develop-
ments and elaborations of the intellectual content of *Being and Time* and therefore as consis-
tent with *Being and Time* and each other. This involves assuming an equivalence between the
Entdeckung (discovering) of *Being and Time* and the *Entbergen* (disclosive looking) of "What Is
Technology?," an equivalence supported by an equivalence between *Entdeckheit* (discovery) and
Unverborgenheit (unconcealment) (1962a, 219). It also involves assuming a rough equivalence
between what is *zuhanden* (ready-to-hand) and *Bestand* (standing reserve). While these seem
reasonable assumptions, it is only fair to say that that is all they are—assumptions. Heidegger
gives us no help in this regard.

2. Citations in this chapter include both the English and German editions of Heidegger's work.

This position has the important consequence that the correspondence theory of truth must be discarded. According to this theory, a proposition is true when it matches the way the world is. According to Heidegger, truth is a consequence of seeing; we bring it to light not through a process of matching, but through a process of revelation: "To say that a statement is *true* means that it discovers the beings in themselves. It asserts, it shows, it lets beings 'be seen' (apophansis) in their discoveredness. The *being true (truth)* of the statement must be understood as *discovering*. Thus truth by no means has the structure of an agreement between knowing and the object in the sense of a correspondence of one being (subject) with another (object)" (1972, 219; 1962a, 201). Since, for Heidegger, truth is revelation (*Entdecktheit*), "to say that an assertion 'is true' signifies that it reveals the entity as it is in itself" (*in seiner Entdecktheit*).

There is a second reason for choosing Heidegger's philosophy for the investigation of scientific visuals, one indissolubly linked to the first: it does not privilege propositions. This is a consequence of *Dasein*, Being's constant existential encounter with the world. Rojcewicz's explanation of the existential nature of this key Heideggerian term cannot be bettered:

> What is most decisive . . . in Heidegger's understanding of humans as *Dasein* is the precise meaning of the "there." The exact sense in which humans are called upon to be the place of a self-revelation of Being. This sense of "there" (as also of *da* in German) is expressed very nearly in a colloquial use of the word in a context admittedly quite foreign to the present one. In the interpersonal domain, a parent may promise a child, or a lover a beloved, to "be there" always for her or him. That is of course not a promise simply to remain at a certain place in space. Nor, at the other extreme, is it a claim of domination. Instead, it is a promise to be available in a supportive way; it is an offer of constant advocacy and nurture. (Rojcewicz 2006, 6)

Heidegger regards science as an ongoing activity whose existential encounters with the world are necessarily the primary focus of philosophical investigation. In *Being and Time*, he says that

> assertion is not the primary "locus" of truth. *On the contrary*, whether as a mode in which uncoveredness is appropriated or as a way of Being-in-the-world, assertion is grounded in *Dasein*'s uncovering [*im Entdecken*], or rather its *disclosedness*. The most primordial "truth" is the "locus"

of assertion; it is the ontological condition for the possibility that assertions can either conceal or reveal the truth—can either be true or false.
(1972, 226; 1962a, 269; translation modified; Heidegger's emphasis)

Not only are theoretical statements derivative of prior unconcealments, but
interpretation exists even in the absence of words: "Interpretation is car-
ried out primordially not in a theoretical statement but in an action of cir-
cumspective concern [umsichtig-besorgenden]—laying aside an unsuitable
tool, or exchanging it, 'without wasting words.' From the fact that words
are absent, it may not be concluded that interpretation is absent" (1972,
218; 1962a, 261; see also 1972, 157; 1962a, 200). Indeed, all communication, including scientific communication, is a communion simultaneously
with others and with the objects of our concern: "when we are explicitly
hearing the discourse of another, we proximately understand what is said,
or—to put it more exactly—we are already with him, in advance, alongside
whatever it is the discourse is about" (1972, 164; 1962a, 207; translation
corrected).

Science, however, is a specific way of looking at the world. In "The
era of the world picture," Heidegger remarks that "the fundamental event
of the modern age is the conquest of the world as picture (als Bild) (1950,
87; 1938, 134), a "structured image (Gebilde)" (1950, 87; 1938, 134). This
"does not mean a picture of the world but the world conceived and grasped
as a picture" (als Bild begriffen; 1950, 82; 1938, 129; emphasis added), one
that permits us to see the world in "its entirety . . . as a system." What is
the source of this view? The modern age itself. It and its keystone, modern science, were inaugurated by two nearly coincident moments in the
early seventeenth century, a philosophical moment, beginning with René
Descartes and culminating with Immanuel Kant, and its scientific counterpart, beginning with Galileo Galilei and culminating with Isaac Newton.
Heidegger explains this Cartesian-Galilean transformation in terms of his
distinction between the "present-at-hand" (vorhanden), the system of concepts available to thought, and the "ready-to-hand" (zuhanden), whatever is
available for use. Science is what is present-at-hand. What is ready-to-hand
is its twin, technology. We are not here concerned with the well-known
fact that technology has its own history, separate from science; we are concerned only with the marriage of science and technology that characterizes
modern science, what Latour calls "technoscience" (1987, 174–76).

It is Descartes who engineers this revolution in our attitude toward the
world; it is he who shifts the guarantee of its objectivity from our existential

encounter with it to an experience wholly within the self. It is this shift, as we shall soon see, that accounts for the startling eruption of the adverb "mathematically" in the second sentence of this passage:

> Until Descartes every thing present-at-hand for itself was a "subject" but now the "I" becomes the special subject, that with regard to which all the remaining things first determine themselves as such. Because— mathematically—they first receive their thingness only through the founding relation to the highest principle and its "subject" (I), they are essentially such as stand as something else in relation to the "subject," which lie over against it as *objectum*. The things themselves become "objects." (1962b, 81–82; 1935–36, 105)

For Heidegger, this paradoxical guarantee of objectivity by means of subjectivity defines modern science, which "is supposed to be based on experience. Instead, it has such a law [as Newton's first law of motion, which posits not real but unrealizable conditions] at its apex. This law speaks of a thing that does not exist. It demands a fundamental representation of things which contradict the ordinary" (1962b, 69; 1935–36, 89).

But the philosophical heirs of Descartes inherited a central puzzle: the link between mind and the world lacked a satisfactory explanation. For Heidegger, therefore, the philosophical moment of modern science culminates not with Descartes but with Kant. According to Kant, since neither pure reason nor its purely empirical counterpart can account for our knowledge of the everyday world or its derivative, the world as understood by science, we must have within us a set of categories and intuitions that permit us to see the world as extended in space and time and causally determined, that is, "mathematical" in the broad sense:

> The mathematical is that "about" things we really already know. Therefore we do not first get it out of things, but, in a certain way, we bring it already with us. From this we can now understand why, for instance, number is something mathematical. We see three chairs and say that there are three. What "three" is three chairs do not tell us, nor three apples, three cats nor any three things. Moreover, we can count three things only if we already know "three." In this grasping the number three as such, we only expressly recognize something which, in some way, we already have. This recognition is genuine learning. (1962b, 57; 1935–36, 74)

Genuine knowledge, then, becomes a consequence of the interaction between the world and inherent categories of understanding, intuitions of space and time. This is Kant's "synthetic *a priori*," the source of what is scientific about our scientific knowledge:

> The mathematical, in the sense of what is movable in space, belongs first of all to the definition of the thingness of the thing. If the possibility of the thing is to be metaphysically grasped [*begriffen*], there is need for such principles in which this mathematical character of the natural body is grounded. For this reason, one group of the principles of pure understanding is called "the mathematical principles." This designation does not mean that the principles themselves are mathematical in the strict sense, but that they concern the mathematical character of natural bodies, the metaphysical principles that lay the ground of that character. (1962, 148–49; 1935–36, 190–91; translation modified)

In this broad sense, the sense that the world is grasped as a picture, the intellectual movement from Galen to Vesalius is mathematical: to both Galen and Vesalius, the body is an elaborate machine. In this sense also, a geological section of the earth's strata by Charles Lyell and a line graph of astronomical velocity and distance by Edwin Hubble are equally mathematical— though Heidegger himself never expressed any concern over such material manifestations of his philosophy.

For Heidegger, visualization of the world as an object guaranteed by the Cartesian subject and open to mathematics and measurement is at the center of modern science: "that the world becomes picture is one and the same event with the event of man's becoming *subiectum* [subject] in the midst of that which is" (1950, 85; 1938, 132). It is a picture whose frame differs from the ordinary in that it *determines* the meaning of the picture it contains. It is the "En-framing" (*Ge-stell*) of everyday experience that creates a rigorously articulated system of efficient causes and effects arrayed in space and time:

> Modern science's way of representing pursues and entraps nature as a calculable nexus of forces [*als einem berechtenbaren Kräftezusammenhang*]. Modern physics is not experimental physics because it applies apparatus to the questioning of nature. Rather the reverse is true. Because physics, indeed already pure theory, sets nature up to exhibit itself [*darzustellen*] as a coherence of forces calculable in advance, it therefore

orders its experiments precisely for the purpose of asking whether and how nature reports itself when set up in this way. (1954b, 29; 1953a, 21; translation modified)

In this connection, Heidegger quotes physicist Werner Heisenberg: "The representing belonging to modern physics is also bent on 'being able to write one single fundamental equation from which the properties of all elementary particles, and therewith the behavior of all matter whatever, follow'" (1954a, 61; 1953a, 172).

This En-framing creates not only modern science, but also modern technology, intellectual Siamese twins, science seeing the world as object present-at-hand, technology seeing it as ready-to-hand, as a standing-reserve, open to use (*Bestand*). Heidegger concludes that modern science and technology represent a particular attitude toward Being, one of exploitation. This conclusion is not our concern; we are concerned only with the extent to which modern technology creates the instrumentation on which the experimentation and observation of modern science rests.[3] For Heidegger "modern physics, as experimental, is dependent on technical apparatus and upon progress in the building of apparatus" (1954b, 22; 1953a, 14). Moreover, observation is itself a technology, given Heidegger's definition of science as a "theory of the real": to do science is "to look at something attentively, to look it over, to view it closely." It means "to look attentively on the outward appearance wherein what becomes present becomes visible and through such sight—seeing—to linger with it" (1954a, 52; 1953b, 163; translation slightly modified). Scientific instruments also include the human eye and, by analogy, the other four senses.

On occasion, En-framing leads to formulations allegedly incapable of being visualized, a frame apparently without a picture, as in quantum mechanics. While apparently true, this allegation is essentially false:

If modern physics must resign itself ever increasingly to the fact that its realm of representation remains inscrutable and incapable of being visualized, this resignation is not dictated by any committee of researchers. It is challenged forth by the rule of En-framing, which demands that nature be orderable as standing-reserve [*Bestand*]. Hence, physics, in all its retreating from the representation turned only towards objects that have alone been standard until recently, will never be able to renounce this one thing: that nature reports itself in some way or other that is

3. Galison's *Image and Logic* (1997) is a brilliant confirmation of this thesis.

identifiable through calculation and that it remains orderable as a system of information. (1954b, 30; 1953a, 23)

In modern physics, it would seem "causality is shrinking into a reporting—a reporting challenged forth—of standing-reserves that must be guaranteed either simultaneously or in sequence. To this shrinking would correspond the process of growing resignation that Heisenberg's lecture [on the philosophical consequences of his indeterminacy principle] depicts in so impressive a manner" (1954b, 30–31; 1953a, 23). Yet, "even where, as in modern physics, theory—for essential reasons—necessarily becomes the opposite of direct viewing, its aim is to make atoms exhibit themselves for sensory perception, even if this self-exhibiting of elementary particles happens only very indirectly and in a way that technically involves a multiplicity of intermediaries (Compare the Wilson cloud chamber, the Geiger counter, the free balloon flights to confirm and identify mesons)" (1954a, 62; 1953b, 173). Despite this limitation, quantum physics is a consequence of the same En-framing as classical physics: "classical physics maintains that nature may be unequivocally and completely calculated in advance, whereas atomic physics admits only of the guaranteeing of an objective coherence that has a statistical character" (1954a, 61; 1953b, 172). Despite this, the invisible world of quantum mechanics can be rendered visible without recourse to intermediaries: Heidegger was not aware of Feynman diagrams, which depict not the paths of subatomic particles but their probabilities.

Although physics is Heidegger's exemplary science, his conclusions apply generally. Each science, of course, has its own character, its own regional ontology: "taking the particular sciences as examples, we attend specifically to whatever is the case regarding the ordering—in any given instance—of the objectness belonging to the object-area" (1954a, 59–60; 1953b, 171). The life sciences are an example; they are also Heideggerian sciences, their late start attributable to the difficulties inherent in imaging life itself as the object of science and to the absence of a tradition of mathematization. Even so, the modern life sciences fall short of the completeness possible in physics: "We feel that what zoology and botany investigate concerning animals and plants and how to investigate it may be correct. But are they still animals and plants? Are they not machines duly prepared beforehand?" (1962b, 31; 1935–36, 41). We will have much more to say on this topic in chapter 3.

While Heidegger's philosophy of science is firmly grounded in our apprehension of the phenomenal world, and permanently anchored in our experiential encounter with that world, his is a philosophical, not a textual

hermeneutics. As a first step in transforming his philosophical hermeneutics into its textual equivalent, we analyze the fundamental perceptual and
cognitive differences between the verbal from the visual; then we characterize Allan Paivio's dual coding theory, a theory that hinges on these differences and strikes us as one possible psychological realization of Heidegger's
philosophical views.

SECOND COMPONENT: VERBAL-VISUAL INTERACTION AND DUAL CODING THEORY

Although the words and images that constitute scientific texts are both
visible and therefore both visual, profound differences exist in these two
communicative modalities. While we see images, we do not ordinarily see
words; rather, we see through them to their underlying concepts: to learn
to read is to learn systematically to ignore the shapes of the letters in the
interest of communicative efficiency. The verbal and the visual also differ
in the way they are organized. Words are ordered in sequential hierarchical
structures composed of combinations of smaller units linked by systematic internal connections. We call this a grammar. There is no grammar
of images. Unlike words, images are ordered into synchronous hierarchies
or nested sets. A face is composed of a nested set of eyes, eyebrows, nose,
mouth, teeth, ears, brow, and cheeks. When organized into larger units,
moreover, words never entirely lose their separate identities; the components of images, on the other hand, tend to lose their separate identities as
they become embedded or nested. Initially at least, we see a face, not its
individual components. Moreover, although the components of images are
images, the components of words are not words. For example, on dollar bills,
"parts of the regions that denote George Washington also denote parts of
George Washington, say his nose, but parts of his name do not denote parts
of him" (Malinas 2003, 257).

The verbal and the visual are also processed differently. Words are processed sequentially; in contrast, images are processed not only sequentially
but also in parallel. Unlike words, images are subject to epistemically relevant spatial transformations. For instance, we may project three-dimensional
objects onto two-dimensional surfaces; we do so when we create a terrestrial map. One image, moreover, may be meaningfully superimposed on another, an effect achieved when lines of latitude or longitude are applied to
maps. In addition, a sequence of visuals may be animated; this is how temporal progression is routinely represented in films and cartoons (Clark and
Paivio 1991, 151–52). Finally, images differ from words in semiotic valence.

Words exist always in the realm of the symbolic: the word "cow" stands for a large milk-producing animal often found grazing on farmland. Images, on the other hand, can be alternately *symbols* standing for some aspect of the world, *icons* representing the world, or *indices* pointing to causal relationships in the world. Because of these differences, in the sciences we would expect a division of communicative, semiotic, and epistemic labor between words and images. Moreover, we expect that these differences would impact the central concern of scientific communication: the arguments that scientists make in favor of the claims they promote.

An important corollary of the differences between words and images is that there can be no scientific arguments the components of which are exclusively visual. To say that there are no visual arguments is not to say that there is no class of scientific problems that can be solved without recourse to language. Our example is the visual operation of rotation. This operation comes into play when we are asked whether a lowercase *p* is the same as an upside-down *b* or *d* (Kirby and Kosslyn 1992, 79). To argue, however, is to do something different; it is to enter the realm of the propositional, to make claims and to provide evidence for those claims. If that is so, any assertion that arguments can be exclusively visual must be mistaken. Evidence, of course, need not be propositional. A photograph can count as evidence; it is truth-bearing (Perini 2005a, 2005b). But for an argument to be an argument, the photograph must be evidence for a claim. And claims must be propositional: they must assert of some state of affairs that it is either true or false, either more or less probable. For example, "when we affirm or deny the proposition *The moon is nearer to the earth than the sun*, neither the moon alone nor the earth, nor the sun, nor the spatial distance between them is the proposition. The proposition is the relation asserted to hold between them" (Cohen and Nagel 1934, 27–29). A picture can certainly illustrate the relationships among moon, earth, and sun; but it cannot assert a proposition either about these entities or about their relationship. A proposition must consist of terms and their predicates: *the moon, the earth, and the sun* are terms; *is nearer than* is a predicate. A picture of the earth is not a term; the distance between the earth and the sun is not a predicate.

This argument against exclusively visual arguments has been seconded by two cognitive psychologists, Oestermeier and Hesse (2000). They point out that epistemic predicates such as "I believe," argumentative predicates such as "I agree because," causal predicates such as "is an effect of," and logical operators such as "all" and "if" cannot be successfully visualized (2000, 94). They infer that "verbal comments and explanations are integral parts of visual forms of communication . . . and the interplay of visualizations and speech acts has to be addressed by all relevant theories of mediated causal cognition" (2000, 75). This seems exactly right. It implies that the analysis of scientific arguments that incorporate the visual must be grounded in a theory that takes seriously the marked differences between the visual and the verbal. Unfortunately, these psychologists also argue that, nevertheless, some arguments *are* wholly visual. In doing so, they confuse two semiotic categories, the iconic and the symbolic. A picture of an arrow is an icon, a recognizable representation of the arrow. We can thus ask how *accurate* that representation is. A picture of an arrow in a technical drawing of a gear (their figure 4), on the other hand, is a symbol standing for the predicate "turns in the direction of." We thus can ask whether it is *true* that the gear turns in that direction. Only propositions can be true or false.

That there are no visual arguments does not mean that visuals are not vital components of scientific argumentation. Their importance stems from a central fact about the visual: while images are not propositional, they can contain more information than can be contained by any set of propositions that can be derived from them. This is information$_1$, the elements of a structured optical field open to interpretation; the interpretations themselves constitute information$_2$ (Heelan 1983, 137, 141). Art historian William M. Ivins Jr. (1969, 57) makes the nature of information$_1$ clear:

> The moment that anyone seriously tries to describe an object carefully and accurately in words, his attempt takes the form of an interminably long and prolix rigmarole that few persons have either the patience or the intelligence to understand. A serious attempt to describe even a simple piece of machinery, such, let us say, as a kitchen can-opener with several moving parts, results in a morass of words that only a highly trained patent lawyer can cope with, and yet the can-opener is simplicity itself as compared to the shape of such a thing as a human hand or face.

Larkin and Simon (1987) see this difference as crucial to scientific problem-solving: "the fundamental difference between our diagrammatic [visual] and sentential [verbal] representations is that the diagrammatic representation preserves explicitly the information$_1$ about the topological and geometrical relations among the components of the problem, while the sentential representation does not" (66).

Chandrasekaran and Narayanan (1993) show us how images can be fully integrated into arguments. In their example, we infer that a child playing in front of a window will be hurt by a rock about to be thrown against it.

Concerning the reasoning exhibited in this case, Chandrasekaran and Narayanan say that

> the first inference about the rock hitting the window pane was made by scanning the scene along a predicted trajectory. Thus, this reasoning involved perceptual and motor operations applied to the environment (eye movement for scanning) as well as computations on an internal representation of the environment (predicting a trajectory). The second inference about the effects of a collision between the rock and the window pane was done by an internal visualization guided by our experiential knowledge. Thus this reasoning involved imaginal operations applied to an internal representation of the environment. The third inference about the child being hurt was made by applying conceptual knowledge to information derived from the visualization. (69, 70)

By computing the rock's trajectory, we infer that it will hit the window-pane, an inference made by means of internal visualization. In turn, this internal visualization provides the context that constrains our next step: an inference about the effects of the collision. To make this inference, internal visualization combines with experiential knowledge: because we know that glass shatters, we infer that the windowpane will shatter. In a final inferential step, we integrate conceptual knowledge. Because we know that flying glass shards will damage human flesh, we infer the effect of the shattering on the child. In this case, images function not as propositions, but as integral, truth-bearing components in arguments that necessarily contain conceptual elements.

This use of images is commonplace in scientific argument. For example, it was on account of images employed in this manner that geologists became convinced that seafloor spreading was real (Giere 1996). It was an image of the double helical structure of DNA, not a proposition, that formed the penultimate step in Watson and Crick's deductive inferential chain (1953a). Their concluding step, the biological implications of the model (1953b), was a verbal inference constrained by the information derivable from this image. These examples imply a four-stage exegetical model: we *perceive* an object or event, we *identify* its structures and components, we *interpret* them, we *integrate* these interpretations into an argument and/or narrative that supports a scientific claim.

Having characterized scientific argument as a consequence of verbal-visual interaction, we proceed to our next step: the characterization of Allan

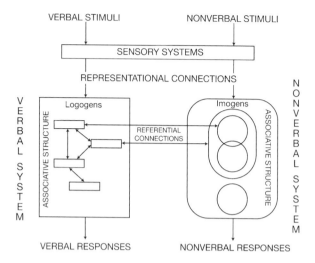

Figure 1.1. A model of verbal and visual processing according to dual coding theory (Paivio 2007, 34). This diagram shows the representational units and their referential (between system) and associative (within system) interconnections.

Paivio's dual coding theory (DCT). Figure 1.1 is a schematic that embodies DCT, a theory that takes the differences between verbal and visual cognition seriously into account. At the top of the diagram, verbal and nonverbal stimuli enter the sensory systems and, by means of representational connections, are channeled into either the verbal or the nonverbal associative system, two separate processing units linked by referential connections.[4] At the bottom of the diagram are the verbal and nonverbal responses that are the output of the model. By means of associative connections in varying configurations and of varying strengths, the system as a whole represents, from one perspective, memory, and, from another, the organization of knowledge in the brain.

In texts that combine the verbal and the visual, meaning is the consequence of the interaction of the verbal and the visual. Figure 1.2 makes this interaction clear. Row by row, the same verbal message is relayed, but the man's differing expressions alter the meaning of the message. In the first

4. The picture is more complex than this simple model reveals. For instance, in the control of action, two visual streams, the dorsal and the ventral, work together "in the production of purposive behavior—one system to select the goal object from the visual array, the other to carry out the required metrical computations for goal-directed action" (Milner and Goodale 2006, 232).

Figure 1.2. Cartoon illustrating interaction between words and pictures (Eisner 1994).
Copyright 1985 by Will Eisner. Copyright 2008 by Will Eisner Studios. Used by
permission of W. W. Norton & Co.

panel of the first row, for example, the message conveyed is puzzlement or consternation, while in the penultimate panel it is smug satisfaction—he is not *really* sorry. Column by column, different verbal messages are conveyed while displaying the same expression. In the penultimate column, for example, a smiling "Goodby" indicates relief at your departure, while a smiling "Blow the works" indicates approval.

One of Röntgen's first X-ray photographs, reproduced as figure 1.3, illustrates the way DCT works in the case of a scientific image. The photograph and its accompanying legend, "Hand mit Ringen" (handwritten on figure

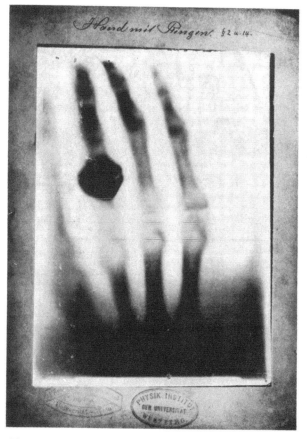

—Picture of a hand which Röntgen sent to the Vienna physicist, F. Exner. The original plate is in the Deutsche Museum, in Munich, and according to a note by Röntgen, was made on December 22, 1895.

Figure 1.3. Röntgen's X-ray photograph of human hand with ring.

top), register first on our perceptual apparatus. Then each is stored separately: the image of the skeletal hand in the nonverbal associative system, the legend in its verbal counterpart.

Images and words are not stored *as* images and words: the brain is a network of neurons and their synaptic connections, not a filing cabinet. Paivio says we store the verbal as logogens, the nonverbal—shapes, sounds, actions, and visceral sensations related to emotions—as imagens; he needs these two neologisms "to distinguish the 'dormant' verbal and nonverbal representational units from their consciously experienced verbal and nonverbal images and their behavioral expressions." While imagens are analogous to what they represent, this is not true of logogens, which are stored in various modalities: "auditory, visual, motor, haptic [i.e., those related to the sense of touch]" (37). Unless they are activated, a process that "occurs via pathways that connect representational units to the external world and to each other," isolated logogens and imagens do not mean. Meaning involves "the juxtaposition of a logogen and referentially related imagens or associatively related logogens" (2007, 41). To discover meaning is to be prompted by two questions regarding Röntgen's photograph: What is it? What does it depict? We answer that *it* is an early Röntgen X-ray, and that it *depicts* a skeletal hand with rings. To provide these answers, we have accessed both associative systems, the verbal and the nonverbal; the referential connections between the two systems made this verbal response possible.[5]

THIRD COMPONENT: ENHANCED DUAL CODING THEORY

While DCT provides us with an appropriate framework for visual and verbal exegesis grounded in the constraints and affordances of human perception and cognition, it is a theory of perception and cognition, not of texts. Because DCT must be supplemented to perform the exegetical tasks we require of it, we have created enhanced dual coding theory (EDCT). According to EDCT, understanding scientific visuals entails four perceptual and cognitive processes:

5. Kress and van Leeuwen (2001) outline an alternative multimodal theory with a semiotic base. Theirs is a grand theory at the highest level of generality. It is meant to apply to all sense data and to every sort of discourse. We think this leap to grand theory is premature; instead, we opt for a middle-level theory with a more modest aim: ours deals only with only two modalities—sight and hearing—and only one sort of discourse, scientific communication. Some may find even this explanatory scope too ambitious.

1. structures and components are perceived as Gestalt patterns;
2. these structures and components are identified and explored by means of regimes of scanning and matching, also a process of perception but one involving consciously directed attention;
3. these structures and components are interpreted by means of Peirce's semiotics (icon, symbol, and index), a strictly cognitive process; and lastly,
4. these interpretations are integrated into semiotic wholes by means of argumentative and narrative structures, a strictly cognitive process.

Different readers, of course, process scientific visuals differently, depending on their stored memories and their knowledge of the subject matter (Carswell et al. 1998; Friel et al. 2001), as well as their level of interest in the overall argument or narrative in which the visuals appear. Despite these differences, research has shown that important task-oriented commonalities exist, commonalities our model attempts to embody.

How do we perceive structures and their components within any given visual representation? To answer that question, we turn to Gestalt psychology. Wolfgang Köhler and his followers were certainly mistaken when they hypothesized that perceptual processes were actually organized along Gestalt lines, hardwired into the brain (Köhler 1947). Given the current state of our knowledge of such processes, it would probably be best to regard the Gestalt "laws of organization [as] opportunistic guides to the viewer as to what will afford desired visual information," and to support the view "that they probably vary widely in level, speed, and power" (Hochberg 1998, 291). It is also worth noting that Gestalt perception has a social and subjective dimension. Faced with unfamiliar patterns, onlookers will continue to be puzzled. For example, the untutored can discern no meaningful patterns in cloud and bubble chamber photographs; for such naïve onlookers, these photographs may as well be examples of abstract art. Further, even knowledgeable readers can interpret the same pattern in different ways, as figure 1.4 illustrates, or even badly misinterpret a pattern, as Edward Tufte (1997) illustrated with regard to a central graph that figured in the space shuttle *Challenger* accident.

On one plausible accounting, there are six basic Gestalt principles. According to *figure-ground*, we see objects automatically as shaped, framed against a shapeless background, one that may, in fact, also have a shape, though we do not perceive it as such. When this background does have a shape—as in the case of the cell structure of tables and the coordinates of

Figure 1.4. The duck-rabbit illusion.

line graphs—we can attend alternately to these structures and to the foreground of bars, data points, and lines of central tendency. When there are alternative "foregrounds," as in the case of lines of central tendency drawn through clusters of data points, we attend either to the lines or the points, placing one or another in the foreground of our perception. Well-known optical illusions discussed in the Gestalt literature—such as the duck-rabbit illusion depicted in figure 1.4 (see Hanson 1958, 10–14)—depend on alternating foreground and background.

We see a second Gestalt principle in operation, *good continuation*, when we follow a line of central tendency, locate a number in a table with the aid of its grid, or complete in our mind's eye the rectangular shape of a whiteboard despite the fact that a man standing in front of it is partially blocking our view. Scientific tables are characterized by a third Gestalt principle, *enclosure*; on the other hand, relationships among their cells are highlighted by means of the fourth principle, *similarity and contrast*. A fifth principle, *proximity*, groups adjoining letters of the alphabet into words even in the case of an early Greek manuscript in which there is no actual separation (Pinker 1990, 84; Hochberg 1998, 260–61; see also Pinker 1983). *Prägnanz* is a final principle, our sense of the whole: we see not a nose, ears, and a mouth, but a face. Röntgen's X-ray photograph (fig. 1.3) exemplifies Gestalt principles. We see not isolated areas of black, white, and gray, but a pattern we will recognize as a skeletal hand. We see the shape of this hand, standing out against a neutral background, an enclosed pattern that exhibits good continuation. Although we can focus on the ring or the individual finger joints, we at first take in the whole. This is the principle of *Prägnanz.*

The structures and components of images perceived as Gestalts are revealed by scanning and matching. The nature of any visual search conducted by these means varies with its goal (Rayner 2009). Galpin and Underwood (2005) and Carpenter and Shah (1998) conducted a series of experiments that supports this generalization. In Galpin and Underwood, subjects were asked to "spot the difference" between two nearly identical visual arrays. Figure 1.5 illustrates the scanning and matching involved in one trial in this series. In the figure, eye movements ("saccades") are represented by lines, eye fixations by circles: the bigger the circle, the longer the fixation. Visible is a pattern of long saccades and progressively longer fixations continuing until the solution is reached—a mirror-image Snoopy toy. In this case, scanning and matching can be purely perceptual.

In contrast to Galpin and Underwood (2005), Carpenter and Shah (1998) are interested in a visual search in which scanning and matching is followed by interpretation. Fig. 1.6 plots the vocabulary scores of twenty- and

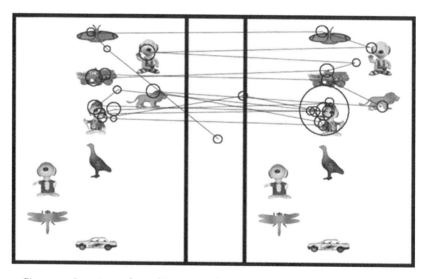

Figure 1.5. Scanning and matching pattern for comparative visual search (Galpin and Underwood 2005, 1325).

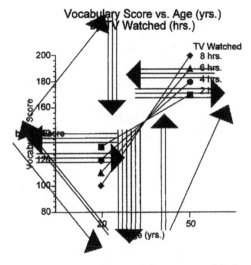

Figure 1.6. Search pattern for a line graph (Carpenter and Shah 1998, 86).

fifty-year-olds as a function of daily time spent watching television and also superimposes on the graph the saccades and fixations from one of the authors' experimental trials. Of this trial, Carpenter and Shah say that

> [this is a] diagram of the number of transitions between regions from a prototypical trial. Each transition is indicated by a line beginning from the center of the regions that were fixated. Note the large number of transitions between the graphic pattern and the x-axis, y-axis, and z-labels regions. This emergent pattern of gazes is consistent with the proposal that the information about the pattern itself is integrated with the interpretation of labels, scales, and units. The viewer, like most, focused on the main x-y functions, saying, "This is vocabulary score versus age by TV watched in hours. And it shows that vocabulary scores increase with age very dramatically for someone who watches a lot of TV and not so dramatically for someone who watches a little of TV." (1998, 86)

Carpenter and Shah assert that "a major component of the interpretation process [in figures and tables] is relating graphic features to their referents and interpreting the referents from different parts of the display" (1998, 96). To do so, readers must first identify their graphic features by scanning and matching. They must then scan and match among the structures and components within the visual, and between the verbal text and

visual, to determine meaningful relationships. For example, they must scan
and match to determine a data component in a particular cell in a table or to
estimate the magnitude of a particular component in a line graph. In scanning, the eyes locate a particular data element or feature, or some aggregate
of data elements or features; in matching, the reader determines meaning by
ascertaining the vertical and horizontal position of data components within
a labeled superstructure. To fully determine meaning, readers may also scan
and match from the data components to the verbal text in the graph or table
itself, to the title, or to the nearby text. Worth noting is that matching does
not involve only the semiotic components of the words or pictures on a
printed page or computer screen. Kosslyn (2002, 273) makes it clear that
"during the process of visual recognition, input is compared with stored visual memories, and if a match is found the stimulus is recognized."

To this principle of scanning and matching, Röntgen's X-ray photograph
constitutes an exception because it is an *existential* image; it attests the
existence of a phenomenon. Figure 1.7, a diagram from Pitman and Heirtzler's (1966) important article on seafloor spreading as a cause of continental

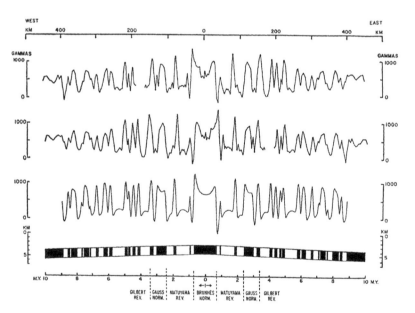

Figure 1.7. Example of graph requiring extensive scanning and matching (Pitman and
Heirtzler 1966, 1166). Magnetic anomalies of the Pacific-Antarctic seafloor are compared
with each other and with the reversals in the earth's magnetic field over geologic time.
Reprinted with permission from the American Association for the Advancement of
Science (AAAS).

drift, illustrates the scanning and matching principles that apply generally to scientific images.[6]

When we scan and match within a complex graph like figure 1.7, we begin in the realm of perception; we link potentially meaningful units together, prior to and in conjunction with interpretation. When we scan downward, for example, we notice that the three jagged lines are remarkably similar, that they very nearly match. We also note that they are matched to two scales, labeled "GAMMAS" and "KM." Scanning downward still further, we see a Gestalt contrast between these jagged lines and the bar of black-and-white boxes at the bottom of the diagram with its verbal labels, and its "M.Y." and "KM" scales. But we also see a match between the ups and downs of the jagged lines and the alternation of white and black boxes. At this point, by scanning and matching, we have identified the significant features of the diagram, though we do not yet know what that significance is. Until prior professional knowledge is applied, we do not know that gammas are a measure of magnetic anomalies, that "M.Y." are millions of years, or that "KM" are kilometers. We do not yet know, but will learn from the text that the first two magnetic profiles are mirror images of each other. Their bilateral symmetry is striking, and, as Pitman and Heirtzler argue, it constitutes strong evidence for evenly distributed seafloor spreading on both sides of an active ridge. We do not yet know, but will learn, that these magnetic anomalies correlate impressively with the periodic reversals in the earth's magnetic field over millions of years, represented by the boxes at the foot of the diagram, another bit of impressive evidence that Pitman and Heirtzler bring forward to support their claim that seafloor spreading originated from the zero point on the distance scale across the graph top, and that therefore continental drift is real.

Visual patterns perceived by scanning and matching or Gestalt principles become meaningful only insofar as they participate in a system of value-laden differences. In effect, one Gestalt principle, that of contrast, forms the perceptual base for differences that are potentially value laden, candidates for meaningfulness generated by the referential connections between the imagens of the nonverbal system and the logogens of the verbal system in DCT (fig. 1.1). Saussure articulates this necessary step to meaningfulness within the verbal system:

> In all these cases what we find, instead of *ideas* given in advance, are *values* emanating from a linguistic system. If we say that these values

6. We also discuss this graph in chapter 5.

correspond to certain concepts, it must be understood that the concepts in question are purely differential. That is to say they are concepts defined not positively, in terms of their content, but negatively by contrast with other items in the same system. What characterizes each most exactly is being whatever the others are not. (1986, 115; Saussure's emphasis)

Saussure's principle may be generalized to apply to any semiotic system, verbal or nonverbal. For example, traffic signals and electrical wiring diagrams are nonverbal systems that also rely for their interpretation on value-laden differences.

No matter how intricate or well organized, no visual pattern has either reference or sense; patterns do not mean. Patterns are invested with meaning only when they are interpreted as one of three categories of Peircian signs (1955, 104–15): icon, index, and symbol. An *icon* is a sign that depicts. A photograph or a drawing of a microbe is an icon. An *index* is a sign whose relation to its object or event is causal or indicative. Geiger counter readings are indices because they are causally linked to the external world, and a "low barometer with a moist air is an index of rain" (Peirce 1955, 109). However, an arrow in a photograph pointing to a microbe is merely indicative. Let us call the latter category of signs "deictic."

There are two sorts of *symbols*: the completely arbitrary symbols of natural and artificial languages, for example, and symbols that are arbitrary only in part. These latter may have religious significance, such as the cross and the Star of David; cultural significance, such as the American flag or the bald eagle; or literary significance, such as Moby Dick or Vanity Fair (Hariman and Lucaites 2002). Peirce (1955, 114) explains the relationship among his three categories and their interpreter:

> The icon has no dynamical connection with the object it represents; it simply happens that its qualities resemble those of that object, and excite analogous sensations in the mind for which it is a likeness. But it really stands unconnected with them. The index is physically connected with its object; they make an organic pair, but the interpreting mind has nothing to do with the connection, except remarking it, after it is established. The symbol is connected with its object by virtue of the idea of the symbol-using mind, without which no such connection would exist.

In EDCT, patterns of perception made potentially meaningful by means of the generalization of Saussure's principles can be interpreted according to

Peirce's semiotic categories. Within a particular category of sign, semiotic transformations are also possible. For example, a photograph of an eye may be used to construct a drawing of that eye, a shift from one iconic mode to another. Other transformations involve a shift from one category of sign to another. For example, the iconic may also be transformed into symbolic: photographs of *an* eye may be converted into a diagram of *the* eye and its various working parts. The iconic may also be transformed into the indexical: a chest X-ray, an iconic image of the human lung, may reveal to a physician the cause of a persistent cough, making the image indexical.

Thomas Sebeok (1976, 1433n) makes an essential cautionary point about the plasticity of Peirce's categories:

> In general, it is . . . inane to ask whether any given subject "is," or is represented by, an icon, an index, or a symbol, for all signs are situated in a complex network of syntagmatic and paradigmatic contrasts and oppositions, i.e., simultaneously participate in a text as well as a system; it is their position at a particular moment that will determine the predominance of the aspect in focus.

As Gérard Deledalle points out: "we must insist . . . on the functional character of these distinctions: what is an index in one semiosis may be a symbol in another. Take, for instance, the symptom of an illness. . . . If this symptom is referred to in a lecture on medicine as always characterizing a certain illness, the symptom is a symbol. If the doctor encounters it while he is examining a patient, the symptom is an index of an illness" (2000, 19–20). Röntgen's X-ray photograph exemplifies this principle. It is certainly iconic of a particular human hand. For Röntgen's scientific peers, however, it was an index of the power of this newly discovered form of light. For us, it is symbolic of a major step forward in science.

This understanding of Peirce's taxonomy as crucially dependent on context is consonant with the views of Nelson Goodman (1972, 445) on meaning making, a position worth quoting at length:

> Comparative judgments of similarity often require not merely selection of relevant properties but a weighing of their relative importance, and variation in both relevance and importance can be rapid and enormous. Consider the baggage at an airport check-in station. The spectator may notice shape, size, color, material, and even make of luggage; the pilot is more concerned with weight, and the passenger with destination and ownership. Which pieces of luggage are more alike than others depends

not only upon what properties they share, but upon who makes the comparison, and when. Or suppose we have three glasses, the first two filled with colorless liquid, the third with a bright red liquid. I might be likely to say the first two are more like each other than either is like the third. But it happens that the first glass is filled with water and the third with water colored by a drop of vegetable dye, while the second is filled with hydrochloric acid—and I am thirsty. Circumstances alter similarities.

The contention of Sebeok, Deledalle, and Goodman that context is central to semiotic interpretation is a generalization of Saussure's principle that meaning is founded on difference.

Understanding the images appearing within scientific texts involves not only interpreting the meaning behind the pattern within a given visual, but also assimilating that meaning into an argument or narrative. It is only by the assimilation into an argument that the X-ray images appearing in Röntgen's 1896 *Nature* article become indices of a new form of radiation; it is only by their assimilation into argument and then into a narrative that Pitman and Heirltzler's diagram becomes an index of the geological history of the earth.

CONCLUSION

This chapter puts into a theoretical context the verbal-visual interactions that communicate meaning in the sciences. We begin with Martin Heidegger's philosophy of science, a philosophy emphasizing visualization. But philosophy will get us only so far in our understanding of scientific visuals; we also need the resources of cognitive psychology. We find Heidegger's philosophical intent realized in Allan Paivio's dual coding theory, a psychological theory that treats the verbal and the visual as two separate but equally important cognitive streams. To adapt DCT to our exegetical purpose, we enrich it, borrowing from Gestalt psychology and scanning-and-matching theory, from Peirce's semiotics of visual communication, and from argument and narrative theory, the subjects of later chapters. We call our synthesis enhanced dual coding theory, or EDCT.

Note that we do not see ours as a general theory of visuals, a theory that applies indifferently to all instances of verbal-visual interaction. For example, it would not apply to arenas of verbal-visual interaction such as theater or Internet video. And there are other limitations to EDCT. These limitations characterize any theory generated in "pre-paradigmatic" times, rife with competing speculations and programs of research far short of completion. Naturally, we would prefer a true theory or, if the philosophical solecism may be permitted, a theory closer to the truth. In broad outline, it is not hard to see what such a theory would look like. At the point in the progress of their science when such a theory has been formed, cognitive psychologists will have arrived at a durable consensus concerning the way human beings perceive, process, and generate meaning through verbal-visual interaction. Moreover, the insights of Gestalt psychology, scanning-and-matching theory, Peirce's semiotics, and argument and narrative theory will have been incorporated into a grand synthesis. At the same time, a philosopher of Heidegger's intellectual stature will have provided an epistemological and ontological justification for this synthesis.

CHAPTER TWO

Understanding Scientific Visuals
and Tables: A Taxonomy

The first step in wisdom is to know the things themselves; this notion consists in having a true idea of the objects; objects are distinguished and known by classifying them methodically and giving them appropriate names. Therefore, classification and name-giving will be the foundation of our science.

—Carl Linnaeus, *Systema Naturae* (1735)

Our task is to apply our model of verbal-visual interaction to scientific tables and visuals. To proceed we need a system of classification with which to differentiate the extraordinary diversity of scientific visuals and tables that now exist. Of course, we already have a system of classification; what is in question is its adequacy: terms like *table* and *diagram*, hallowed by common usage and of considerable practical use, prove on inspection to be problematic when rigorously applied. For example, we might define *table* as "an arrangement of data in rows and columns, often including verbal headings explaining what the data represent." The primary purposes of such a visual are data presentation and retrieval. But if this is so, is the periodic table of chemical elements a table? Figure 2.1 reproduces one of the first versions of Mendeleev's periodic table. The table is indexical, that is, it points to the calculable nexus of forces at the atomic level that creates the distinctive entities we call "elements," organizing these into six columns, each according to its atomic weight. The columns are the periods, so arranged that a new row begins when its elements begin to display chemical properties analogous to those of another period. Fluorine (F = 19), chlorine (Cl = 35.5), bromine (Br = 80), and iodine (J = 127) form a group of this sort. Mendeleev was able to forecast that gaps in this table would be filled by the discovery of "many new elements" with predictable properties (1869,

I.	II.	III.	IV.	V.	VI.*)
			Ti = 50	Zr = 90	? = 180
			V = 51	Nb = 94	Ta = 182
			Cr = 52	Mo = 96	W = 186
			Mn = 55	Rh = 104,4	Pt = 197,4
			Fe = 56	Ru = 104,4	Ir = 198
		Ni =	Co = 59	Pd = 106,6	Os = 199
H = 1			Cu = 63,4	Ag = 108	Hg = 200
	Be = 9,4	Mg = 24	Zn = 65,2	Cd = 112	
	B = 11	Al = 27,4	? = 68	Ur = 116	Au = 197 ?
	C = 12	Si = 28	? = 70	Sn = 118	
	N = 14	P = 31	As = 75	Sb = 122	Bi = 210
	O = 16	S = 32	Se = 79,4	Te = 128?	
	F = 19	Cl = 35,5	Br = 80	J = 127	
Li = 7	Na = 23	K = 39	Rb = 85,4	Cs = 133	Tl = 204
		Ca = 40	Sr = 87,6	Ba = 137	Pb = 207
		? = 45	Ce = 92		
		? Er = 56	La = 94		
		? Yt = 60	Di = 95		
		? In = 75,6	Th = 118 ?		

Figure 2.1. Mendeleev's first periodic table (1869, 406).

406). Two were soon discovered: gallium (Ga = 68), an analogue of aluminum in the boron family; and germanium (Ge = 70), an analogue of silicon in the carbon family.[1]

In the periodic table, we are witness for the first time to an ordering of the elements based on the order of nature. In fact, it *just happens* that arranging the elements by their atomic weights, an arrangement that reveals something essential of their nature, is tabular in form because nature *is just that way*. In Mendeleev's hands, the table has been transformed from a representation in which space is a means of locating data points into one in which space stands for relationships among the elements. Some of these, such as atomic number, ionization energy, and electron negativity were, in Mendeleev's time, yet to be discovered. The periodic table, it turned out, was knowledge that generated knowledge. The periodic table is no more a table than a whale is a fish.

Is a trellis diagram a table? Figure 2.2 reproduces a trellis diagram that appeared in an article by R. A. Fisher and E. B. Ford (1947), a study of a small and isolated population of a large day-flying moth with scarlet hind wings, *Panaxia dominula*. The study was these scientists' entry into a controversy with Sewell Wright. The gist of the dispute can be easily stated:

1. The actual atomic weights differ from those predicted: Ga = 70 and Ge = 73. Also note that in this early table, ordering by chemical properties trumped atomic weight for several elements, e.g., tellurium (Te = 128) and bismuth (Bi = 210).

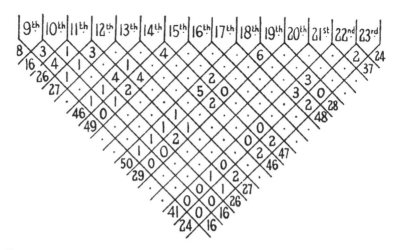

Figure 2.2. Example of table with variant structure: Moth captures and releases (Fisher and Ford 1947, 155). Reproduced by permission of Nature Publishing Group.

did evolutionary change occur by slight selection pressures on large populations, as Fisher and Ford believed, or by chance mutations occurring in small populations, as Wright did?

How is this configuration to be read? Scanning downward and to the right from the 14th of July, we reach the number 46, the number of specimens of *Panaxia dominula* released on that date. Scanning downward and to the right, also from the 14th, we reach the number 5; scanning upward and rightward from this 5, we end at July 18th. Of the forty-six moths released on the 14th, then, five were recaptured on that date. Finally, following the diagonal for the 18th downward to the left, we discover that the sample captured on that date was fifty moths. Scanning downward and to the right from July 18th, we discover that of that fifty, forty-eight were released. This innovative matrix makes visible, at a glance, key aspects of the research method Fisher and Ford employed. The trellis diagram is designed to retrieve quantitative interrelationships that would defeat the organization of a standard table. But while not structured as an array of horizontal rows and vertical columns, this trellis diagram is a table because, like a table, its primary purpose is data presentation and retrieval.

But what are we to say of the trellis diagram in figure 2.3 of B. C. Coull's "Species Diversity and Faunal Affinities in Meiobenthic Copepoda in the Deep Sea"? It appears in an article about the distribution of this class of bottom-feeding crustaceans along the continental shelf, the continental slope, and the lower slope and abyss—a total distance seaward of about 500 miles. A bold diagonal line divides the square into two right triangles, creating a hybrid scientific visual. The top triangle forms a table that allows us to retrieve individual data points by scanning and matching; for example, depths at 14 and 17 meters share 19 percent of their species. But it is not the numbers in the upper table that dominate this figure; it is the shaded boxes in the bottom triangle, derived from those numbers and arranged in Gestalt patterns that make trends in the data immediately clear: for example, "a sharp faunal break at the shelf-slope boundary [500 meters], the

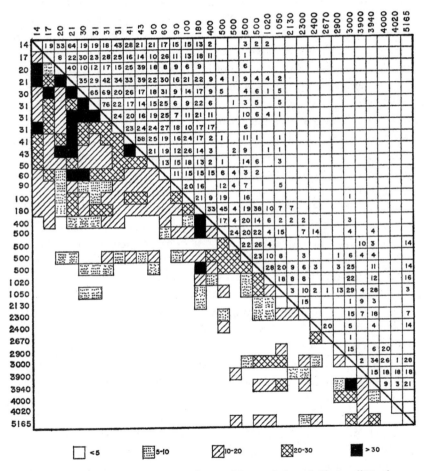

Figure 2.3. Trellis diagram showing degree of Copepoda faunal affinity off North Carolina coast (Coull 1972, 50). Axes: Ocean depth in meters (5165 meters = 3.2 miles); boxes display percentage of species shared.

area of greatest temperature change" (Coull 1972, 50), as indicated by the
sparsity of filled squares below 500 meters. This is a visual that combines
the virtues of a table and a graph, those of a graph predominating. At the
same time, as it is not a typical graph, it does not plot data points or curves
on an x-y coordinate system.

These examples illustrate the problem that arises from using for analytical purposes such common designators as *table* or *graph*. Properly understood, these examples also illustrate Ludwig Wittgenstein's views on
categorization (1953), views to which Eleanor Rosch long ago gave psychological and experimental substance (Rosch 1973, 1975, 1999; Rosch et al.
1976). To Rosch, as to Wittgenstein, concepts cannot be captured by formal
definitions. Games are Wittgenstein's example. Soccer and solitaire are both
games; on this there is universal agreement. But what formal definition
could capture the united essence of both? Among the types of tables and
graphs, just as among games, although family resemblances exist, conceptual overlap among members does not mean that every member shares with
all of its relatives a common set of criterial attributes. It means, rather,
that in each conceptual family, there are prototypical members, those "with
most attributes in common with other members of that category and . . .
with least attributes in common with other categories" (Rosch and Mervis
1975, 576). As we move away from the prototype, the amount of overlap decreases; for example, when we think of a chair, we are more likely to picture
a dining room chair than a beanbag chair. In a beanbag chair, gone are the
legs, back, and seat; all that is left is function. One person can sit down in
a beanbag chair—sort of. Analogously, when we think of a scientific table
we are more likely to think of Hubble's prototypical table, illustrated in
figure 2.4 (discussed later), than Mendeleev's table, illustrated in figure 2.1.
In Mendeleev's table, unlike Hubble's, gone are the central functions, data
presentation and retrieval. All that is left are the rows and columns. Rosch
and Mervis conjecture that it is our tendency to think in terms of prototypes rather than in gradients that creates the illusion that criterial definitions are possible (1975, 583). Rosch's point (1999, 66) about the nature of
categories is general:

> Gradients of membership judgments apply to the most diverse kinds of
> categories: perceptual categories such as *red*, semantic categories such
> as *furniture*, biological categories such as *woman*, social categories such
> as *occupations*, political categories such as *democracy*, formal categories
> that have classical definitions such as *odd number*, and *ad hoc*, goal-
> derived categories such as *things to take out of the house in a fire*.

Such judgments are, of course, fallible. While whales are not fish, they share certain characteristics with them, so much so that they can be mistaken for fish. While the periodic table is not a table, it shares certain characteristics with tables, so much so that it can be mistaken for a table.

In our view, the classification of visual representations must proceed along lines compatible with Rosch's insights. The system we will employ is based on a single organizing principle, the overriding purpose a visual representation serves. We classify representations into seven categories: 1) those designed to present and retrieve data, such as prototypical tables, 2) those designed to express data trends, such as prototypical line and bar graphs, 3) those designed to express spatial relationships, such as prototypical maps and molecular models in chemistry, 4) those designed to express space-time interrelationships, such as prototypical geometric diagrams representing the movement of objects over time, 5) those in which space is a metaphor for a relationship other than spatial, such as prototypical flowcharts, process diagrams, and circuit diagrams, 6) images whose purpose is virtual witnessing of what the scientist saw in the field, laboratory, office, or observatory, such as prototypical photographs of star clusters or cloud chamber events, and 7) prototypical images designed to reveal the function of equipment.[2] Does a map of the globe express spatial or space-time relationships? The first, if it is designed to fix the relative location of Helsinki and London; the second, if it is designed to be a unit in a sequence that illustrates Alfred Wegener's theory of continental drift over vast tracts of time.

Overriding purpose is a criterion that can deal with apparent anomalies. For example, line and bar graphs also use space metaphorically: distance between points stands for quantities or for time intervals. Nevertheless, this use of space and this reference to time are incidental to the overriding purpose of these graphs: to represent data trends. We also recognize that some representations are hybrids. Coull's trellis diagram, for example, incorporates both a tabular representation and a graphic representation of the distribution of crustaceans in space. Nevertheless, it does so only in order to represent trends in their distribution: it is for this reason that we place it in category two, data trends. Finally, we recognize an inclination to collapse categories six and seven, since both depict some object in space. Nonetheless, it is overriding purpose that differentiates the one from the other.

---

2. Various scholars have attempted to construct taxonomies for scientific visuals, including Tufte (2001), Kosslyn (2006), and Desnoyers (2011). See Desnoyers (2011) for an excellent summary of these efforts. Ours differs in its emphasis on underlying purpose as the key factor in determining the appropriate category.

In the remainder of this chapter, we will give examples of each of these categories, analyzed according to the model we have presented in chapter 1. It is a model in which Gestalt principles permit us to perceive at a glance patterns among the structures and components of tables and visuals: scanning and matching, to identify deeper interrelations among the structures and components; and semiosis, to interpret these interrelationships and integrate them into scientific arguments and narratives. The overall purpose of this chapter is to exemplify the utility of both our system of classification and our model for verbal-visual interaction.

REPRESENTATIONS OF DATA: TABLES

The primary purposes of tables are data presentation and retrieval. Their prototypical structure is the real or virtual network of horizontal and vertical lines whose intersection creates cells or data spaces. With tables, viewers must scan and match to determine a data component in a particular cell. As Bertin has observed (1981, 1983), graphs correspond closely to tables: in effect, graphs are tables transformed for the purpose of exhibiting trends. Because of this difference, tables are processed much differently from graphs. In tables, the Gestalt principle of figure-ground permits us to foreground the data and to treat the superstructure as background; the principle of enclosure highlights each data cell. The principles of good continuity and similarity are also involved: the first permits us to move easily among columns and rows, while the second allows us to treat the cells equally, an equality designed to deprive the cells of special meaning.

Figure 2.4 reproduces a table that originally appeared in Edwin Hubble's article (as table 1) on the motion of galaxies in relation to their distance from the Earth. Searching Hubble's table to find a particular piece of information, readers must first survey its surface. Words attract the eye first, imposing order and meaning (Kress and van Leeuwen 1996, 23–24). A typical reader might travel from the title, "Nebulae Whose Distances Have Been Estimated . . . ," to the column headings, and then move from left to right along the top row as if reading a line of text: object (galaxy) name, m_s, r, v, m_t, and M_t. By themselves, these column headings are meaningless. To understand them, readers must drop down to the footnote, which defines them as astronomical parameters. Thus informed, readers are prepared to make sense of the data. They might choose a particular galaxy by scanning down the first column to the last entry, "4649." Then they might scan across the row that entry initiates to match 4649 with its distance (r) and velocity (v).

TABLE 1

NEBULAE WHOSE DISTANCES HAVE BEEN ESTIMATED FROM STARS INVOLVED OR FROM MEAN LUMINOSITIES IN A CLUSTER

OBJECT	m_s	r	v	m_t	M_t
S. Mag.	..	0.032	+ 170	1.5	−16.0
L. Mag.	..	0.034	+ 290	0.5	17.2
N. G. C. 6822	..	0.214	− 130	9.0	12.7
598	..	0.263	− 70	7.0	15.1
221	..	0.275	− 185	8.8	13.4
224	..	0.275	− 220	5.0	17.2
5457	17.0	0.45	+ 200	9.9	13.3
4736	17.3	0.5	+ 290	8.4	15.1
5194	17.3	0.5	+ 270	7.4	16.1
4449	17.8	0.63	+ 200	9.5	14.5
4214	18.3	0.8	+ 300	11.3	13.2
3031	18.5	0.9	− 30	8.3	16.4
3627	18.5	0.9	+ 650	9.1	15.7
4826	18.5	0.9	+ 150	9.0	15.7
5236	18.5	0.9	+ 500	10.4	14.4
1068	18.7	1.0	+ 920	9.1	15.9
5055	19.0	1.1	+ 450	9.6	15.6
7331	19.0	1.1	+ 500	10.4	14.8
4258	19.5	1.4	+ 500	8.7	17.0
4151	20.0	1.7	+ 960	12.0	14.2
4382	..	2.0	+ 500	10.0	16.5
4472	..	2.0	+ 850	8.8	17.7
4486	..	2.0	+ 800	9.7	16.8
4649	..	2.0	+1090	9.5	17.0
Mean					−15.5

m_s = photographic magnitude of brightest stars involved.
r = distance in units of 10^6 parsecs. The first two are Shapley's values.
v = measured velocities in km./sec. N. G. C. 6822, 221, 224 and 5457 are recent determinations by Humason.
m_t = Holetschek's visual magnitude as corrected by Hopmann. The first three objects were not measured by Holetschek, and the values of m_t represent estimates by the author based upon such data as are available.
M_t = total visual absolute magnitude computed from m_t and r.

Figure 2.4. Example of data representation in table format: distances, visual magnitudes, and velocities of galaxies (Hubble 1929, 169).

The product of this matching and scanning would be the proposition, "Galaxy 4649 is 2.0 million parsecs from Earth and is moving at a velocity of 1090 kilometers per second." Tables are not visuals but are a way of arranging a series of semiotic components—verbal or numerical—so that a large number of such parallel propositions can be efficiently generated.

Tables work best at conveying precise values. Although their rows and columns may also contain numerical patterns, such patterns are usually submerged in the busy background of other data, tabular lineation, and words. Such patterns can be perceived, if at all, only at some cognitive expense. Is there a pattern to the data in Hubble's table 1? We need a graph to tell us the answer most efficiently.

REPRESENTATIONS OF DATA TRENDS:
LINE AND BAR GRAPHS

The prototypical structure of line graphs consists of Cartesian coordinates: horizontal and vertical axes on either two or four sides. Their typical components consist of labels defining the object of measurement; tick marks defining the units of measure; dots, circles, or squares symbolizing data points; and solid, dashed, or dotted lines tracing the best fits to data or equations. Since the primary function of line graphs is the display of data trends, Gestalt pattern recognition is the central task needed to comprehend them. The secondary task, the identification of individual data points, proceeds by scanning and matching. In searching a graph to determine individual data points, viewers scan to a particular data position in the graph body and match it to a position on the abscissa and ordinate, sometimes guided by a grid system within the graph. The result is a proposition in the form, "At point x, the value of y is z."

Figure 2.5, from the same article by Edward Hubble, plots the radial velocity versus distance for a series of galaxies, or what Hubble refers to as "extra-galactic nebulae." To identify data points on this graph, viewers scan along the x-axis from left to right and stop at a distance equal to, say, 2 x 10^6 parsecs, then scan up the vertical line and match the solid points on the line with the numbers on the y-axis to the far left. This scanning and matching yields the following proposition: "At 2 x 10^6 parsecs from the earth, the measured galaxy velocity varies from about 500 km/sec to 1100 km/sec." In principle, viewers can derive a large set of such propositions from Hubble's graph by repeated scanning and matching. In fact, Hubble would not have cared whether his readers actually scanned and matched their way to any specific proposition. He was interested in trends.

Graphs do not merely arrange data for viewers to easily derive a large set of propositions like those just mentioned; they mainly answer implied questions about trends (Bertin 1981, 2). In the case of figure 2.5, there is the implied question, is there a correlation between galaxy velocity and distance from the earth? Understanding the answer requires viewers to perform three main tasks: pattern recognition, interpretation of the meaning of that pattern, then integration of that meaning into the argument in which it appears.

In Hubble's line graph, data points and their lines of central tendency are perceived as such according to the Gestalt principle of *Prägnanz*; these components form a single perceptual unit because of their proximity and similarity. At the same time, different sets of these points are separated

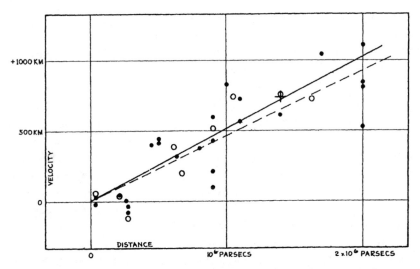

Figure 2.5. Example of data representation in line graph: Galaxy velocity versus distance (Hubble 1929, 172). Radical velocities, corrected for solar motion (from the values in fig. 2.4), are plotted against distances estimated from involved stars and luminosities of nebulae in a cluster. The black circles and solid line represent the solution for solar motion using the nebulae individually; the open circles and broken line represent the solution combining the nebulae into groups; the cross represents the mean velocity corresponding to the mean distance of 22 nebulae whose distances could not be estimated individually.

visually by contrast, the reverse of this last principle. Gestalt patterns allow viewers to perceive that the data points steadily inch upward, more or less in synch with the two slanting lines. This trend indicates that though there is considerable scatter among the points, as a rule galaxy velocity increases linearly with distance from the earth.

In Hubble's line graph, as well as in line graphs in general, knowledge-able viewers can interpret data trends almost immediately, "without apparent mental effort" (Cleveland and McGill 1985, 828). This is because the patterns are usually quite simple, and no barrier exists that interferes with perception. A central tendency can remain flat, increase or decrease, follow some easily recognized shape like a U or an S, or fluctuate regularly (Pinker 1990, 109). Of course, in some cases there is no central tendency at all: the data can vary unpredictably.

That there is a linear correlation is only part of Hubble's message. The Gestalt principle of contrast alerts readers to differences among the symbols (closed and open circles) and lines (solid and dashed), differences that cannot be interpreted without verbal instruction. These Hubble provides in the figure caption: solid circles represent the distance-velocity data for the twenty-four galaxies listed in a table; open circles, nine data components in which the distances of nearby galaxies are grouped to improve measurement accuracy; the solid line, the best-fit line to the solid circles; and dashed line, the best-fit line to the open circles. Note also the cross in the middle. It represents the mean velocity corresponding to the mean distance of twenty-two additional nebulae. The proximity of the data components to the two lines drawn through them suggests a more nuanced version of the general claim that there exists a "linear correlation between distances and velocities." The solid line relies upon many data points with a wide range of scatter; the dashed line has fewer data points but far less scatter. Both lines yield the same interpretation, though there is a difference in the strength of the claim that can be made on their basis.

The data in Hubble's graph are symbolic; they stand for galaxies in motion. Integrated into the article as a whole, the data are indexical: they are evidence for an argument concerning a previously unknown linear correlation between galaxy velocity and distance. It is an argument with narrative implications for the history and future of the universe.

Bar graphs are another means for illustrating data trends and making comparisons. Similar to line graphs, they consist of superstructures (their axes) and data elements (their bars along with labels). The superstructure of bar graphs differs from that of line graphs in that the horizontal axis consists of entities or processes rather than quantities. As with line graphs, readers perceive the superstructure according to the Gestalt principles of figure-ground and good continuation; they perceive the data elements according to the Gestalt principles of enclosure, proximity, and similarity. But while enclosure clearly differentiates the bars from their background, they can be differentiated from each other by any of three graphic variables of contrast: color, texture, or shading (Bertin 1983, 96). Readers shift from data components to superstructures and back by means of the Gestalt principle of figure-ground; this same principle permits readers to foreground each set of bars against their vertical and horizontal axes.

The bar graph in figure 2.6 quantifies the changes in the degree of eye disease in patients with hyperthyroidism treated with one of the following: radioiodine alone, radioiodine and prednisone, or methimazole alone. Having perceived the graph's patterns according to Gestalt principles, viewers interpret its bars by scanning across from one to another and matching heights. The axes of the underlying structure are interpreted by means of their labels: an ordinal scale on the vertical axis, a classification by modality of treatment on the horizontal axis. In figure 2.6, the data elements are

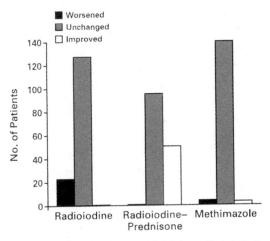

Changes in the Degree of Ophthalmopathy in Patients with Hyperthyroidism Who Were Treated with Radioiodine, Radioiodine and Prednisone, or Methimazole.

Patients in whom ophthalmopathy developed are included in the group with worsening. The determination of patients' status was based on an overall evaluation of ocular changes, variations in the ophthalmopathy-activity score, and the patients' own evaluation, as described in the Methods section.

Figure 2.6. Example of data representation in bar graph: Effect of different drug regimes on ophtalmography of patients with hyperthyroidism (Bartalena et al. 1998, 76). Reprinted with permission of Massachusetts Medical Society.

the length of the bars and their contrasting grayscale values. The meaning of this length is fixed by the axis labels; the meaning of the contrasting grayscale values, by the key on the upper left. Scanning and matching of each bar produces an interpretation of the form, "For treatment x, y number of patients were in condition z." More important, each set of bars generates a statement concerning the efficacy of a treatment modality, an efficacy determined by visual comparison. For example, this level of interpretation tells us that radioiodine treatment—represented by the first set of three bars— worsened the conditions of some patients, made no difference to most, and improved the condition of none. There is a higher level of interpretation, the meaning of the graph as a whole. This level demonstrates that only treatment with radioiodine and prednisone improved patients' condition. The bar graph must also be integrated into the article of which it is a part. This graph constitutes the evidence for the central claim that "of the three treatments, only the administration of prednisone after radioiodine therapy was associated with amelioration of preexisting eye disease during the one-year follow-up period" (Bartalena et al. 1998, 78).

The data in the bar graph are symbolic; they stand for relative number of patients cured. Integrated into the article as a whole, the data are indexical: they are evidence for an argument for the relative efficacy of treatments.

In some respects, graphs exceed tables in cognitive and communicative power. In graphs, though not in tables, parallel processing is possible (Dubin 2002, 25–30); simultaneously, we take in such varied perceptual parameters as shape, shading, intensity, spatial position, and color. It is this parallel processing that enables readers to see the graph as a whole, virtually instantaneously, a single conceptual message in visual form (Pinker 1983, 5; Pinker 1990, 120–121; Bertin 1983, 146). The superiority of graphs over tables, however, is relative to the task at hand. When individual data are of primary importance, tables are the appropriate choice; when relationships among data are primary, graphs are (Kosslyn 2006, 30–33).

REPRESENTATIONS OF ARRANGEMENTS IN SPACE: MAPS AND MOLECULAR MODELS

We now move from representing data and data trends by means of spatial relationships to representing objects and events actually distributed in space; we move to such representations as maps and molecular models. There are many kinds of maps: geologic, geographic, and astronomical. They are united in purpose, serving "as a source of information or an aid to decision making and behavior in space" (MacEachren 2004, 12). We scan and match to arrive at such propositions as "The distance between Chicago and New York is 720 miles," or "Venus orbits the Sun between Mercury and Earth." While maps have many practical and informational uses, our concern here is for maps that present spatial information in the interest of scientific argument.

The typical map's underlying structure is analogous but not identical to that of data graphs: a coordinate system quantifies actual rather than Cartesian space. Semiotic components vary widely depending on the map's purpose. For a geographic map, a common component is the scale bar for equivalent distance, that is, "the ratio between the size of the map and the size of the piece of the environment it is trying to show" (Karrow 2007, 3). For such maps, other prototypical semiotic components are place-name labels and shapes representing locations such as cities or mountains or coastlines or national borders. Color also often conveys meaning: dark blue for highways, light blue for water, green for forests, brown for desert, and so on. A color scale with differing intensities of hue can communicate quantitative information such as average temperature, height above sea level, and population density.

We will take as our prototype a relatively simple map in the Fisher-Ford moth study (1947) from which we extracted the trellis diagram illustrated in figure 2.2. The map in figure 2.7 locates the site of their field experiment. It was vital to this study that the research area be isolated, that it form, in effect, a laboratory in a local wooded park. The map's code is quasi-representational. Employed are stylized representations of paths, streams with arrows indicating direction of flow, marsh grass, and deciduous and coniferous trees. This code makes it easy actually to see the isolation required: *Prägnanz* permits us to view as one area a marsh surrounded on all sides by either woods or agricultural land. Both are "totally unsuited to *P. dominula*, which is strictly confined to the marsh" (147). These moths—their

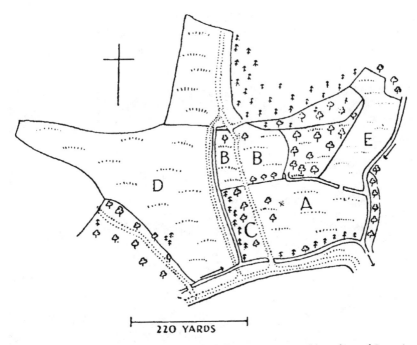

Figure 2.7. Example of spatial relationships: Map of areas occupied by colony of *Panaxia dominula*. "The species is most abundant in the western section of A and throughout B, also among the trees in the open places in C up to the conifers. Specimens are also to be found sparingly along the eastern border of D, only occasionally straying to the middle of it. They are rather scarce in area E. . . . Except where other indications are given, the site is bounded by agricultural land" (Fisher and Ford 1947, 146). Reproduced by permission of Nature Publishing Group.

wingspans are nearly two inches across—exist in this area in a typical form and in two varieties, one of which is *medionigra*. It is a comparison between the shifting populations of *P. dominula* and its variant, *medionigra*, over eight seasons that is at the core of Fisher and Ford's study. A central releasing point, marked with a deictic *x* in the A section of the map, ensured that captured insects mingled randomly with the population as a whole. Integrated into the article's argument, the map symbolizes the isolation required to generate credible results uncontaminated by other colonies of moths.

Scientific maps also frequently superimpose an analytical scheme on the spatial relationships they depict. The color-coded global maps we reproduce as figure 2.8 illustrate the tumultuous geodynamics all the way to the Earth's core. In this figure, six globes are foregrounded against a dark background and arrayed in tabular fashion: two rows and three columns. A network of white lines divides the globes up according to longitude and latitude. The globes themselves represent the earth as viewed from outer space at a distance of 35,000 kilometers or 22,000 miles. The Atlantic side is the focus of visual attention in the top row; the Pacific side, in the bottom.

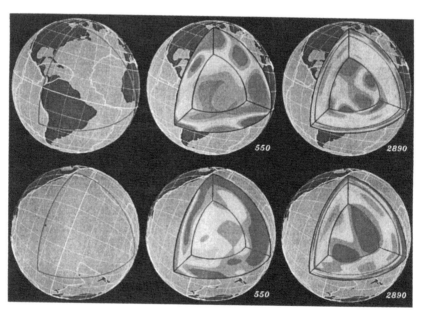

Figure 2.8. Example of geologic map: Three-dimensional views of the earth's interior below the Atlantic (top row) and Pacific (bottom row) (Dziewonski and Woodhouse 1987, 46). In the original color image, yellow lines mark plate boundaries; white lines, longitude and latitude; red lines, boundary of interior being modeled. Colors within wedges are explained in our text. Reprinted with permission from AAAS.

The first column shows the earth's surface; in the original figure the irregular yellow lines trace the tectonic plates and the red lines, the triangular wedges featured in the adjacent columns. As the labels state, the triangular wedge in the second column excavates 550 kilometers (342 miles) below the ocean surface; in the third column, 2890 kilometers (1796 miles), roughly halfway to the Earth's center. The authors selected these distances for a reason. The wedge illustrated in the second column traverses the Earth's upper mantle; that in the third column maps down to the point where the Earth's mantle meets the core.

Without any verbal context, readers immediately recognize the globes as planet Earth by the Gestalt principle of *Prägnanz*; the six globes also form a single perceptual unit because of their proximity and similarity. At the same time, readers immediately recognize differences among the globes as they scan and match right to left, and top to bottom. Key to understanding the meaning behind the differences is a verbal explanation of the color code within the wedges: shades of blue indicate cool matter sinking downward; shades of red indicate hot material bubbling upward. Thus informed, readers can almost immediately interpret the maps' argument. There is considerable geodynamic activity within the Earth's mantle. Moreover, that activity differs significantly in the Atlantic and Pacific wedges: the first is characterized by more of the cooler, sinking matter, the second, by more of the hotter, rising matter. The yellow lines allow the authors (and readers) to scan and match for rough correlations between the outlines of the tectonic plates and the geodynamics below them. The six globes are both icons of the earth and indexes of its geodynamic activity all the way to the core-mantle boundary.

A very different type of map is the molecular model. Historically, molecular models have been as vital to chemical and biochemical communications as mathematical equations are to theoretical physicists. They translate measurements of a compound's constituents, structure, and chemical bonding into a visual form that approximates the structure at an atomic scale (Hoffmann 1991). While there are many variations (Goodsell 2005), the typical schematic is simplicity itself. The basic semiotic components are circles or sets of abbreviations signifying different elements or compounds (H for hydrogen, C for carbon, O for oxygen, Ru for ruthenium), and lines indicating the bonding of one or more components to another. The resulting structures simulate three-dimensional patterns in two-dimensional space. In figure 2.9, for example, the bond diagram of $HRu_6(CO)_{18}$ is a specific realization of the generalized octahedral structure depicted just below it.

In the bond diagram, wedges and different line thicknesses create the illusion of three dimensions, foregrounding some elements over others. The

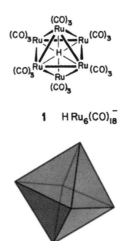

1 $H Ru_6(CO)_{18}^-$

Figure 2.9. Example of modecular model: Bond diagram of the octohedral structure of the organometallic complex HRu6(CO)18 (Hoffmann 1981, 995). Reprinted with permission from AAAS. Octahedron reproduced below for comparison.

Gestalt principle of good continuation allows for easy navigation. As with graphs and tables, we perceive these structures by pattern recognition and identify their components by scanning and matching. For those trained in interpreting such schematics, recognition as a particular geometric structure is virtually immediate. Scanning and matching from one bond to the next, viewers identify the details behind the pattern. In doing so, they see hydrogen imprisoned in the middle, bonded to six $Ru-(CO)_3$ compounds. They also find one $Ru-(CO)_3$ compound sitting on each of the six vertices in the octahedron. This typical organometallic compound consists of a metallic fragment (Ru) and a ligand (CO). All ligands "must possess an electronic arm by which they attach themselves to the metal. This is their basicity, their donor function, and it consists of one or more pairs of electrons used for bonding to the metal" (Hoffmann 1981). This molecular image is at the center of Hoffmann's argument in favor of a particular structure for organometallic compounds. His iconic model represents all $HRu_6(CO)_{18}$ complexes in nature. More than that, it is indexical: it points to a particular binding mechanism as a causal factor in its structure.

REPRESENTATIONS OF SPACE-TIME RELATIONSHIPS

The two prototypical examples we have chosen to illustrate the visual representation of space-time relationships are Galileo's depictions of the relative

Fig.	Date	East	West
1	Jan. 7	• • O	•
2	8	O • • •	
3	10	• • O	
4	11	• • O	

Figure 2.10. Example of space-time representation: Galileo's first four observations of the relative positions of Jupiter and its satellites (Galileo 1610, 72).

positions of Jupiter and its moons (fig. 2.10) and Charles Lyell's depiction of deep time as revealed by the Earth's strata under England's coast (fig. 2.11). Like the prototypical geographic maps and molecular models just described, Galileo's and Lyell's images are stylized representations of their objects. They are alike in that they represent relationships between space and time. In this respect they differ from the local and global maps and $HRu_6(CO)_{18}$ molecule (figs. 2.7 through 2.9), which represent only space.

Galileo is interested in showing that Jupiter has several moons orbiting the planet. He does this by illustrating the varying positions of these celestial bodies over two months, starting from January 7, 1610. This effort produced a series of sixty-four astronomical pictures. Figure 2.10 reproduces the first four.[3] The figure's superstructure is a network of horizontal and vertical lines whose intersection creates cells. This basic structure is the same as a table. Its purpose, however, is not to retrieve data but to display spatiotemporal relationships: in particular, the positions of Jupiter (circular disk) and three smaller nearby celestial bodies (dots) as observed by Galileo through a telescope on four nights in January.

In reading the accompanying text in *Sidereal Messenger*, we learn that on January 7 Galileo observed what he assumed were three "fixed stars" close to Jupiter, one on the right side and two on the left. On January 8 the "fixed stars" had moved: the three points of light now appeared on the right side of Jupiter. Although Galileo waited with "intense longing" for the next night, he was greatly disappointed because the sky was too cloudy: hence, there was no diagram entry for January 9. The next two nights Galileo found that the three satellites had again changed position: only two were observable, and he assumed the third was hidden behind Jupiter. By January 11, it

3. The diagram reproduced, though that of the translator, is faithful to Galileo's diagrams. Galileo's own diagrams are not reproduced, as the amount of space they take is extravagant and would disrupt the flow of our text.

was as "clear as daylight" to Galileo that there are at least "three [satellites] in the heavens moving about Jupiter as Venus and Mercury around the Sun" (47). On January 13, he observed a fourth satellite.

Scanning and matching determines that the circles and dots in figure 2.10 are Jupiter and its satellites, respectively. The Gestalt principle of contrast differentiates the planet Jupiter from its satellites; it also differentiates the observations of their relative positions over four nights. Until the conclusion of *Sidereal Messenger*, Galileo interprets his observations in terms merely of satellite revolution; at the end, however, he integrates these observations into an argument that includes the whole of the known solar system: "we have a notable and splendid argument to remove the scruples of those who can tolerate the revolution of the planets round the Sun in the Copernican system, yet are so disturbed by the motion of one Moon about the Earth, while both accomplish an orbit of a year's length around the Sun, that they consider that this theory of the constitution of the universe must be upset as impossible" (69).

Figure 2.11 is the central visual component of an argument concerning the remote geological past, when the Ice Ages occurred. It comes from Lyell's *The Antiquity of Man* (1863). In the figure, the Gestalt principle of enclosure differentiates one stratum from another, each identified by scanning and matching. Gestalt contrast allows us to differentiate among stratal features, the representation of which constitutes a code hovering between the iconic and the symbolic. It is this code that forms the perceptual basis for their classification. The open shapes in stratum 1 represent flints; the stippling in stratum 2 represents chalk; the upright shapes in stratum 3 represent tree stumps; the elongated shapes in strata 3' represent the remains of flora and fauna; the filled-in shapes in stratum 4 resemble boulders; the wavy lines in stratum 5 resemble geological contortions; the broad stip-

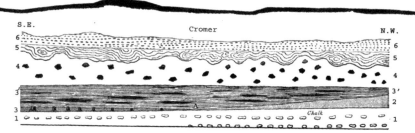

Figure 2.11. Example of space-time representation: succession of the strata in the Norfolk cliffs, extending several miles northwest and northeast of Cromer, England (Lyell 1863, 213).

pling in stratum 6 resembles a mix of sand and gravel. The text makes clear that these six strata are arranged in order in both space and time, the lowest being the oldest. While stratum 5 follows 4, however, strata 2 and 3 are roughly contemporaneous, a forest and a beach that fan out from point A, directly below "Cromer." Time flows downward so that as the numbers on the right and left sides decrease, geologic time increases.

As a whole, figure 2.11 also hovers between the iconic and the symbolic. It is not a depiction of the actual geological configurations of the Norfolk cliffs northwest to northeast of the Cromer jetty; rather, it permits us to infer that in any actual configuration the strata would appear roughly in the order represented and would consist roughly of the features Lyell has assigned to them. It is these features and this order that permits Lyell to infer that after a period of warmth in which trees could grow (levels 3 and 3'), there occurred a period of extreme cold in which "glacial formations" (217) deposited the erratic boulders in level 4 and "ice islands" (222) distorted the stratum in level 5. It is thus that he integrates this figure into his argument for extreme climate change over geological time, the existence of an Ice Age. This argument implies a changed narrative of the history of the earth.

REPRESENTATIONS IN WHICH SPACE IS A METAPHOR: FLOWCHARTS, PROCESS DIAGRAMS, CIRCUIT DIAGRAMS

In *Reading Images*, Kress and van Leeuwen (1996, 101–3) make an important distinction between the "topographical" and "topological," between representations in which space is space, such as terrestrial maps (our third category above), and those in which space is a metaphor for something else. We treat here three kinds of visual representation in this latter category: flowcharts, process diagrams, and circuit diagrams. In flowcharts and process diagrams, space is metaphoric. In this characteristic, they parallel line graphs. But whereas line graphs indicate trends, that is, patterns of data over real time, flowcharts and process diagrams are models of events that *just happen* to take place in time: in these cases, all that matters is the fact of sequence. In circuit diagrams, space is a metaphor as well; it is a metaphor for the functional relationships that exist among the components.

Flowcharts consist of a superstructure of rectangles linked by arrows and arrayed into three patterns—rightward, downward, and branching. Figure-ground permits readers to perceive the superstructure, while good continuation permits easy navigation from arrow to arrow. Enclosure allows readers to see the informational rectangles as units, while similarity makes them equal in importance.

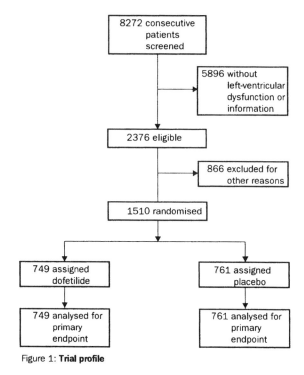

Figure 1: **Trial profile**

Figure 2.12. Example of space as visual metaphor: flowchart of experimental design in medical study (Køber et al. 2000, 2054).

The flowchart shown in figure 2.12 is part of a prospective randomized, double-blind study that compares the effects of the heart medication dofetilide with a placebo in patients with recent myocardial infarction and left-ventricular dysfunction. Flowcharts are read in one of two directions, from left to right as in Western languages or from top to bottom as is traditional Japanese. In each case, the reader must understand from context that moving forward is moving forward in time. In the particular case of figure 2.12, the reader must also understand that arrows are vectors, that right-branching boxes are subtractive, and that parallel boxes represent parallel, temporally simultaneous steps in a medical protocol. In other words, the reader must be familiar with a visual code. With such well-established conventions within the medical community in place, the underlying structure of figure 2.12 can channel the perception necessary for interpretation. Reading downward, medical scientists are able to interpret these rectangular configurations as meaningful elements in a system of patient screening over time. They understand that of 8,272 consecutive patients screened for the study, 5,896 without left ventricular dysfunction or information about such dysfunction were excluded, leaving 2,376. After the exclusion of 866 for other reasons, the remaining 1,510 were assigned randomly to take either dofetilide (749) or a placebo (761).

This flowchart tells readers that a prospective randomized clinical trial is being reported, one that exhibits the appropriate experimental design. The flowchart is, therefore, essential to the article's credibility: experimental design is what makes this science science. Why a flowchart? Why not plain English sentences? Gestalt principles enable readers to see at a glance not just the content, but also the structure of the study's screening and assignment: the flowchart is anchored not only in a network of meaningful sentences, but also in a network of meaningful scientific practices. Integrated into the article's argument as a whole, the flowchart is part of a larger, indexical process: it guarantees that the relative efficacy of treatments has a sound scientific basis.

Process diagrams like figure 2.13 are another means of representing sequencing. In an article published in the *American Journal of Psychiatry*, Hobson and McCarley unveil their alternative to Freud's theory of dreams; their theory, the activation-synthesis theory, was constructed from the vantage of nearly eighty years of subsequent research on brain physiology and dreaming. Figure 2.13 is a visual comparison between Freud's psychoanalytical model of dreaming and the then-new model posited by activation-synthesis theory. Instead of assuming with Freud that dreams represent suppressed wishes, activation-synthesis theory posits the periodic activation of neuronal dream generators during REM (rapid eye movement) sleep, and accounts "for the maintenance of sleep in the face of strong central activation of the brain" by assuming the simultaneous blocking of both sensory input and motor output (1335). The brain stem stimulus is activated "by the perceptual, conceptual, and emotional structures of the forebrain." This is "primarily a synthetic constructive process, rather than a distorting one as Freud presumed" (1347). The partial coherence of some dreams is merely an artifact of the forebrain doing "the best of a bad job" in making some sense out of the "relatively noisy" signals transmitted from the brain stem (1347).

Hobson and McCarley's diagram compares their theory with Freud's by means of three boxes linked with arrows indicating that the flow is left to right. In the diagram, each box is a stage in the dream process, each arrow a vector leading to the next stage (D is the dream state; W is the waking state). Each set of boxes symbolizes a competing theory. The diagram elucidates a key difference between the two theories, made evident by scanning and matching between the two sets of linked boxes: activation-synthesis theory requires no mechanism for disguising the manifest content of dreams. In the upper diagram, this mechanism is represented by a barrier labeled "CENSOR." At this barrier, in Freudian theory, the dream is distorted by dream work, creating its manifest content, that is, what the dreamer remembers. Its latent content, its real meaning, lies in the patient's preconscious, awaiting psychoanalysis. No such barrier exists in the lower diagram. When integrated into their overall argument, Hobson and McCarley's diagram calls into doubt the dream interpretations of psychotherapy. According to their argument, dreams cannot be interpreted in the Freudian sense; they have no meaning beyond their surface appearance.

The superstructure of circuit diagrams consists of lines forming a partial or full rectangle or set of rectangles whose sides are periodically interrupted by symbols standing for electrical properties such as resistance and voltage. Straight lines allow easy navigation as the viewer traces the electrical current from one component to another. Unlike geographic maps but like subway maps, the specific location of a component in a prototypical circuit diagram is not representative of its physical location.

Two Models of the Dream Process*

PSYCHOANALYTIC MODEL

| UNCONSCIOUS Repressed wishes strive constantly and actively for discharge. | EGO Wishes to sleep, withdraws cathexes. Day residue stirs up unconscious wish threatening to disrupt sleep and invade consciousness. | CENSOR | SLEEP | WAKING |

PRECONSCIOUS · · · · · · · · · · · · · · · · · ·

LATENT CONTENT

DREAM WORK Disguises dream thoughts via displacement, symbol formation, pictorialization, condensation, and so forth.

REPORT MANIFEST CONTENT

ACTIVATION-SYNTHESIS MODEL

NONSPECIFIC STATE GENERATOR Sets level of brain's constituent neurons to determine D state.

ACTIVATION of sensory neurons, motor neurons, and "visceral" neurons via disinhibition in D state. The route, intensity, and pattern of activation differ from W state.

SYNTHESIS Integrates disparate sensory, motor, and emotional elements via condensation, displacement, and symbol formation. Increase in intensity gives vividness. Change in pattern gives scene and plot shifts.

REPORT

Figure 2.13. Example of space as metaphor: process diagram comparing Freudian theory and activation-synthesis theory (Hobson and McCarley 1977, 1346). Reprinted with permission from the American Psychiatric Association.

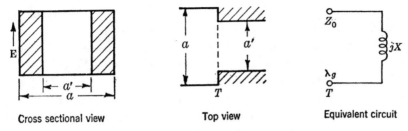

| Cross sectional view | Top view | Equivalent circuit |

Figure 2.14. Example of space as metaphor: cross section and top view of "junction of two rectangular guides of unequal widths but equal heights" and its equivalent circuit (Marcuvitz 1951, 168). As its name suggests, a "waveguide" is simply a physical conduit for directing electromagnetic waves from one location to another. The hatching signifies the waveguide walls.

In figure 2.14, the far-right image is a "microwave analogue of the electrical engineers' more practical representation" (Galison 1997, 821), created to capture the behavior of microwaves passing through waveguides developed as part of radar research during World War II. In this equivalent circuit diagram, space is a metaphor for the functional relationships that exist among the semiotic components along the virtual circuit starting at point T and ending at Z_o; the coiled symbol in the latter is code for an inductor, a storer of magnetic energy. In the left side diagrams, on the other hand, space is just space as in an engineering or architectural drawing. That the two sets of diagrams represent "the same thing" is hinted at by the shared italic capital T. Although in some sense equivalent, the diagrams on the left side clearly contrast with the one on the far right: the former are physical, the latter electrical.

Figure 2.14 (far right image) is one of the simpler of a large set of equivalent circuit diagrams in *Waveguide Handbook* (Marcuvitz 1951), each of which transforms into images microwaves radiating through different configurations of long hollow tubes having various discontinuities. In turn, these images are transformed into mathematical equations, and these equations into calculated results and theoretical understanding.

DRAWINGS AND PHOTOGRAPHS AS VIRTUAL WITNESSES

Realistic drawings and photographs permit us to see what the scientist saw in the field or laboratory. They differ from the other visual forms in that they lack prototypical structures other than the rectangular frame that encloses them; nevertheless, their framing is not casual, as in a snapshot; rather, it is a visual constraint, helping the viewer focus on the scientifically

salient. As Lynch and Woolgar (1990b) point out, realistic drawings and photographs "are more than simple representations of natural order" (5); they
consist of carefully orchestrated scenes in which the viewer's attention is
focused on salient semiotic components that fit into some larger cognitive
structure: "What is presented is presented as already distinct from ordinary
or lived-bodily space precisely by the limited and selected-out 'framing'
of the image presentation" (Ihde 1998, 91). While, as Gary Malinas (2003)
notes, these scenes are "parts of possible worlds"—a fossil site, the night
sky—for scientists, they can be part of only one possible world: the world
that their science posits as an explanatory context.

Realistic drawings of nature dominated the scientific literature until the
last quarter of the nineteenth century; at that point, it became possible to
reproduce photographs in journal articles and books. Up to that point, realistic drawings counted as evidence; after that point, photographs began to
take their evidential place. But because realistic drawings could depict their
subjects more perspicuously, and because in so doing they could teach their
viewers how to locate and interpret what was scientifically significant, they
did not disappear from scientific texts. Writing in 1880, the Leipzig embry-
ologist Wilhelm His (1880, 6) puts this difference well:

> Drawing and photograph are complementary, without replacing one an-
> other. The advantages and disadvantages of a drawing in relation to a
> photograph lie in the subjective elements that are at work in its making.
> In every intelligently executed drawing the essential is consciously sepa-
> rated from the inessential and the connection of the depicted forms is
> shown correctly, according to the draftsman's conception. Accordingly,
> the drawing is more or less an interpretation of the object, involving
> mental effort on the part of the draftsman and embodying this effort for
> the viewer, whereas the photograph reproduces the object with all its
> particularities, including those that are accidental, in a certain sense as
> raw material, but which guarantees absolute fidelity.

Our first example, figure 2.15, is H. B. Whittington's drawing of *Opabinia Regalis*, a marine creature, about four inches long, one that flourished
more than 500 million years ago. Within its virtual frame, *Opabinia Regalis*
is seen from three angles. To the left, we view the creature from the top; to
the right, we see it on its right side. In the center, we see three cross-sections,
whose positions can be located by following the arrows flanking them. Distinguishing the individual constituents requires scanning and matching.
Scanning from top to bottom (a), we see a long proboscis, five eyes, body,

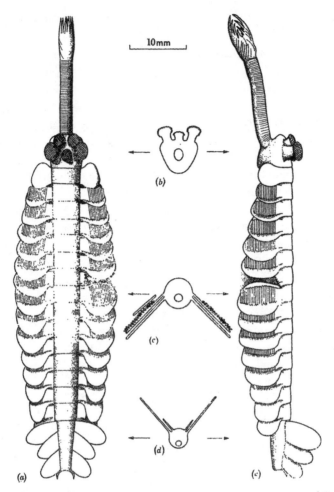

Figure 2.15. Example of realistic drawing of object of study: reconstruction of *Opabinia regalis* based on fossil (Whittington 1975, 34).

and tail. By matching the center structures (b, c, d) with the skeletal ones flanking it (a, e), we can re-create in the mind's eye a three-dimensional image of the lobes, gills, and fins along the body. Furthermore, matching the ten-millimeter scale at the top with the bodily parts, we can determine the size of any segment.

This iconic image of a newly discovered species from the dawn of life is striking on its own terms, yet it is only through verbal-visual interaction that this alien creature from a distant time truly comes to life. Whittington writes that the wormlike trunk visible in a and e "is interpreted as having been muscular, adapting to exploring the [sea-bottom] sediment for food, trapping it, and conveying it to the mouth. . . . No jaw structures are known, so the food was presumably soft" (1975, 39). Such words, combined with the image before us, bring this animal to life in a way that images alone cannot. Readers become virtual witnesses to a prehistoric scene. Furthermore, by integrating Whittington's drawing into the article's overall argument, they transform this creature into a representative of "an ancestral group of segmented animals" related to but different from arthropods (a lobster is an arthropod) and annelids (earthworms are annelids). This image is part of a larger argument and a narrative about the evolution of life on Earth.

Scientific photographs, our next form of realistic representation, first appeared in the mid-nineteenth century (Keller 2008), though not until the next century did photographic reproductions appear routinely in scientific books and articles.[4] Over time, photographs changed scientific visualization as profoundly as did the invention of the graph in the eighteenth century. They have captured in images everything from observations in the field to the path of a subatomic particle to the spiral of the Andromeda galaxy. They are the result of virtually direct contact with the objects of science. Photons create these images directly, or more often by means of a vision-enhancing technology, such as a microscope, telescope, X-ray machine, or cloud chamber. Such images make us all "virtual witnesses" (Shapin and Schaffer 1985, 60) who see exactly what the scientist saw and analyzed.[5]

Figure 2.16 shows a set of three astronomical photographs taken with a ten-foot telescope located atop a mountain in Chile in March 1995. It documents the formation of a supernova, or exploding star, in the most distant reaches of the universe. In accord with our model, viewers detect three squares with differently sized darkened circles scattered within, shapes that stand out against a grey background. Underneath the first square they see "7 Mar"; underneath the second, "30 Mar"; underneath the third, "(30 Mar)–(7 Mar)." In addition, they see two deictic arrows pointing to two circles, one with the label "SN 95K." Confronted by a series of discrete im-

---

4. One of the first significant scientific books amply populated with photographs is Charles Darwin's *The Expression of the Emotions in Man and Animals* (1872).

5. It is a role that the invention of photograph-altering software has somewhat diminished. Fraud is always a possibility, but the technicalities of microscopic image acquisition make unintentional misrepresentation also possible (Rossner and Yamada 2004, North 2006). Furthermore, photographs—and any other sort of visual representation—can on occasion deceive even the trained eye (Galison 1998). Finally, we recognize that photographic objectivity is a contingent matter with its own history (Daston and Galison 2010, 161–73).

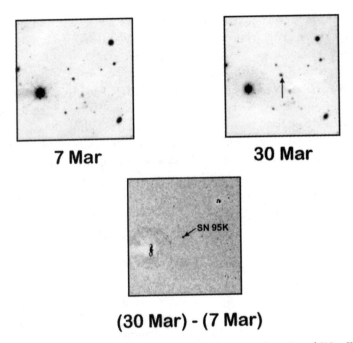

Figure 2.16. Example of photograph: Images demonstrating detection of SN 95K, a new supernova (Schmidt et al. 1998, 55). Reproduced by permission of American Astronomical Society.

ages arranged in rows, they "read" them in the same order as they would straight text: from left to right, and from top to bottom. They begin with the upper left image above "7 Mar," then scan to the upper right image above "30 Mar." Scanning and matching between these two images reveals that their patterns are virtually alike: the only exception seems to be the arrow pointing to one of the dots.

To discover what that arrow betokens, viewers scan down to the image over "(30 Mar)–(7 Mar)." Gestalt comparison reveals a difference between this image and the two above: in this case, the arrow focuses the viewers' attention on the appearance of SN 95K. Without the arrow, without the "SN 95K," the most salient component in this third image would be the largest within the visual field, the dark shape on the left-hand side. Interpretation allows knowledgeable readers to arrive at the following proposition: the luminosity of the small dark circle identified by an arrow has increased substantially over the three-week period in March. This circle represents the first observation of a new supernova (SN), the eleventh (K) such discovery that year (1995). The photographic triptych must now be integrated into the argument of an article, one that concerns a new calculation of the density of matter in the universe based on measurements of SN 95K.

What is this one picture worth? It allows us to infer the existence of a transient astronomical event in the distant past (5,800 million years ago), a task words could never accomplish. Without the picture-taking telescope, that event would have been the equivalent of a tree falling alone in a forest. Subsequent measurements based on the photograph and other photographic images showed that this distant object was moving away from Earth at an astonishing speed. In a larger theoretical context, this image contributed evidence to an argument that the expansion of the universe is accelerating, not decelerating as previously believed. The big bang explanation for the origin of the universe, and the subsequent narrative dependent on it, was in need of some repair.

REPRESENTATIONS OF EQUIPMENT

Key to the rapid advance of science are the creation, adaptation, and modification of research equipment and its integration into an experimental design composed of multiple components—at least one of which is by the mid-twentieth century frequently some image-generating machine. Such representations symbolize the reproducibility of the research results scientists produce. Figure 2.17 comes from a then virtually unknown Michael Faraday. It was published in the *Quarterly Journal of Science* (1821) as part of an addendum to an article reporting his first significant discovery: the reciprocal circular motion of a magnet and an electric current, the first foray in his lifelong enterprise of developing a field theory of magnetism and electricity. In his laboratory diary, and in a subsequent letter to Charles-Gaspard de la Rive, Faraday interpreted this motion as an index, the result of a fundamental force clearly to be differentiated from the only forces previously recognized—Newtonian forces that can act only on bodies and can produce only motion in a straight line.

Gestalt contrast allows us to perceive the components of the apparatus; scanning and matching from the image itself to the descriptive text allows us to identify them. Glass vessels filled with mercury flank both sides of a brass T fixed to a wooden base. On the left, on the application of an electrical current, a cylindrical magnet will circle around a wire; on the right, in contrast, the application of a current will make a wire circle around an embedded cylindrical magnet. In figure 2.18, not only is the external appearance of the apparatus depicted, but so are components and linkages that

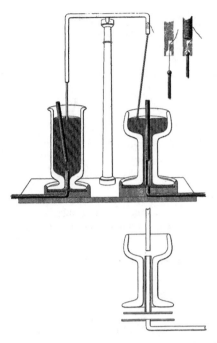

Figure 2.17. Example of drawing of research equipment: Apparatus for demonstrating the reciprocal circular motion of a magnet and an electrical current (Faraday 1821, 806).

would otherwise be hidden from view: these reveal that nowhere in the apparatus's construction is there any barrier to the unmediated realization of the natural effects. Faraday's depictions of his apparatus are iconic. But we are not meant to see his actual equipment as it might appear in the laboratory; we are meant to see an idealized version others can use to produce the same effects. Furthermore, Faraday's depictions are iconic in the interest of indexicality: they allow us to see through his apparatus so that we can see the causal structure of the world. We interpret the meaning of the components of the apparatus properly only when we integrate their functions into the larger purpose of Faraday's grand experimental design, an argument in favor of a new fundamental force, electromagnetism.

Some scientific visuals re-create not experimental apparatus, but an experimental site. To do so, they employ a visual code composed of stylized iconic representations of their research equipment somewhat like a circuit diagram. Figure 2.18 appears in an article demonstrating that a beam of light can be captured in a cloud of liquid sodium cooled to near-absolute-zero and suspended by electromagnetic forces. In this case, meaning is anything but self-evident. To understand this diagram, we must activate the experimental

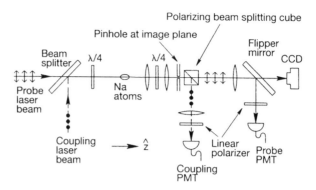

Figure 2.18. Example of experimental arrangement of research equipment: apparatus used in experiment to stop laser beam in a supercooled sodium cloud (Liu et al. 2001, 491). Reproduced by permission of Nature Publishing Group.

arrangement in the mind's eye. It has three main regions, reading from left to right: input to the experimental site, the site itself, and the output. The interpretation of this image emerges from Gestalt pattern recognition and scanning and matching between text and visual, and within the visual itself. On the left are the probe and coupling lasers, whose initial direction differs by ninety degrees. They are combined with the beam splitter, adjusted with a quarter wave plate ($\lambda/4$), and directed into the supercooled sodium cloud ("Na atoms" in figure), which brings the beam to a complete halt for a very brief period. On the right are various devices for detecting and measuring the beam after it exits the sodium cloud, including a digital imaging device (CCD). The visual depicts the relative positions of each segment of the ex-perimental apparatus, its orientation in space, and its interactions with the laser beams that produce the desired effect. The figure allows the reader vi-sually to re-create the experiment in a way that no prose can hope to match. Integrated into the overall argument, this image embodies the source of the knowledge claim encapsulated in the article's title, "Observation of Co-herent Optical Information Storage in an Atomic Medium [sodium cloud] Using Halted Light Pulses."

CONCLUDING REMARKS

The two semiotic modalities scientists routinely employ in concert—the verbal and the visual—have very different histories. The visual realization of the verbal, writing, hardly evolved after its invention: alphabetic script developed slight, though significant, adjustments that improved legibil-ity: word separation, punctuation, paragraphing, and the motivated used of majuscule and miniscule letters and of descenders and ascenders. But scientific visuals evolved in a major way, relentlessly exploiting the opportunities provided by a new ecological niche, the printed page. At the start of the Scientific Revolution, this exploitation was well under way: natural philosophers were already using the vertical dimension of the page to form tables, projecting a spherical surface onto a plane to form maps, and employing perspective to project the third dimension of an image onto a plane. Later, they would use space as a metaphor for time and turn the space of the page into the Cartesian space of line and bar graphs. Our taxonomy has tried to bring order to this explosion of scientific creativity. To do so is, however, to misrepresent an evolving as a static phenomenon. In the following chapter, we try to make sense of this evolution by placing it within the framework of Heidegger's philosophy of science: we try to show not only that scientific visuals evolved, but that they evolved in a certain direction, one that conceived the world as a calculable nexus of forces for which the data-driven graphic is a prototypical realization.

CHAPTER THREE

Visual Evolution and the Heideggerian Transformation

Variation, whatever may be its cause, and however it may be limited, is the essential phenomenon of Evolution. Variation, in fact, *is* Evolution. The readiest way, then, of solving the problem of Evolution is to study the facts of Variation.

—William Bateson, *Materials for the Study of Variation* (1894)

In the third book of *Gulliver's Travels* (1735), Jonathan Swift abandons his tight narrative and indulges forthrightly in broad satire. In his account of the Grand Academy of Lagado—read: the Royal Society of London—he is particularly merciless. Among his targets are the attempts to reform the language. He takes as his particular target Thomas Sprat, first biographer of the Royal Society, and his pronouncement that scientific communication must "return back to the primitive purity and shortness, when men delivered so many things in an equal number of words" (Sprat 1667, vol. 2, xx). His wicked lampoon takes this pronouncement literally, and has as its climax two "sages" exchanging views without the benefit of natural language, both burdened with heavy bags of "things" that they laboriously pass to and fro in order to communicate (190–91).

Swift has been proven right: science has found natural language indispensable. But Swift has also been proven wrong. Artificial languages, such as those of mathematics and logic, though ultimately they depend on natural language for their meaning, *are* more perspicuous, the notation that Leibniz invented for his calculus being a good example. Equally important, science has found that "things" (in the form of scientific visuals) are as indispensable to its communication as is natural language.

The insight is not ours. Nearly a half century ago, Martin Rudwick (1976) argued for the centrality of visuals in geologic literature; more important, he showed that, as the nineteenth century unfolded, geological visuals became increasingly prominent and increasingly theoretical in character, routinely picturing the earth as "spatiotemporal magnitudes of motion,"[1] to borrow a phrase from Martin Heidegger (1938, 119). In Rudwick's view (1976), this transformation in geological visuals was not solely a matter of the conceptual evolution of an individual science. Two more-mundane factors also contributed: the emergence of geology as "a self-conscious new discipline with . . . well established institutional forms (150)," like specialized scientific societies and their periodicals; and advances in imaging technology—"aquatints, wood engravings, steel engravings and lithographs" (151)—that greatly facilitated the integration and reproduction of printed texts with "maps, sections, landscapes and diagrams of other kinds" (150).

We would extend Rudwick's claim from geological science to the sciences in general; we would assert that over the centuries, the visuals of the various sciences alter their character; more and more, they depict not the natural world as an artist might, but as Heidegger phrased it, as a "calculable nexus of forces" (1954b, 29) open to mathematics and measurement. As Heidegger also recognized, this transformation in the picturing of the natural world originated at different times for the different sciences. Geology was not the first. By the early seventeenth century, astronomy and physics had already evolved into "modern sciences" in Heidegger's sense.

The theoretical status of the visual in the verbal-visual semiotic at the heart of scientific communication was not a given. In *The Eye of the Lynx* (2002), art historian David Freedberg explores the unsuccessful efforts of Federico Cesi, Galileo's associate, in depicting the world's flora and fauna. Freedberg shows that Cesi's depiction of these in all their variety was self-defeating. Diagrams illustrating scientifically significant spatial relationships among types were needed: not a means of looking *at* the world, but a means of looking *through* it to its causal structure. The diagram was a vehicle for this new kind of seeing; it "appealed for its effectiveness to order, logic, and reduction. Each of its marks was significant. Take away a line from a diagram, and it is nonsense. Take away a mark from a picture and all that is lost, in the worst of cases, is some part of its affect" (396). Physicist Richard Feynman (1965, 13) generalizes the discovery that Cesi missed: Feynman says that "there is . . . a rhythm and a pattern between the phenomena of nature which is not apparent to the eye, but only to the eye of analysis," a metaphor that is more than a metaphor. Feynman's insight lies in the shadow of Heidegger's philosophy of science.

---

1. Literally, "space-time-motion magnitudes."

According to Heidegger, every age—whether that of ancient Greece, the Middle Ages, or the modern era—is characterized by its own worldview, its *Weltanschauung*. Our age, the modern age, is distinct because it has as one of its "essential characteristics . . . its science" (1938, 116).[2] Because they differ in *Weltanschauung*, modern scientists differ from intellectuals of other eras; their particular task is methodologically to explore a "specific domain of objects, or *Gegenstandsbezirk*." Uniformly, these explorations are implemented by means of a "rigorous research procedure" designed to calculate law-like relationships within this domain "in its future course" or to verify something "about it as past." To commit to this process is to view the world as a "structured image," a *Gebild*, consisting entirely of "spatio-temporal magnitudes of motion"(119), a characterization that explains why Heidegger entitles his essay "The Era of the World Picture [*Weltbild*]." By "world picture," he most emphatically does *not* mean anything so tactile as an actual picture of the world, but rather "the world conceived and grasped as a picture" (129), that is, as a structured image of a specific object domain.

Nonetheless, these structured images, we contend, manifest themselves in the different sciences at different times by means of actual pictures representing the world as spatiotemporal magnitudes, open to either mathematization or, in a later stage, mathematics per se. Just as tourists returning from Vienna share with friends and family snapshots taken while touring its first or second *Bezirk*, chemists and biologists share with colleagues graphs and photographs documenting their visits to their particular *Gegenstandsbezirke*, their peculiar domains of spatiotemporal objects. These snapshots, these graphs and photographs, are eventually transformed into Bruno Latour's "immutable and combinable mobiles" (1986; 1987, 227), mobiles made widely available by means of the invention of printing[3] and, latterly, by the Internet.

While we think that the centrality that Heidegger bestows upon picturing the world gives his philosophy of science an exegetical edge over any analytical version, we do not think the divide between these two philo-

---

2. Throughout, the translation has been modified slightly on occasion, the better to reflect Heidegger's German.

3. According to Elizabeth Eisenstein (1979), "the printed word first assumed importance for the European scientific community in the 1470s when Regiomontanus [Johannes Müller] established the Nuremberg Press" (461). Eisenstein also emphasizes the importance of "printed illustrations" to these early scientific books (587–90), which appeared on separate pages, integrated into the text, or pasted into the margins. The 1485 astronomical book *Sphaera Mundi*, coauthored by Regiomontanus, Johannes de Sacrobosco, and Georg von Peuerbach, even included an illustration printed with three-color ink.

sophical traditions makes them incommensurable. What, after all, is Heidegger's view? That modern science sees the world through the lens of mathematics in the broadest sense, and that its goal is to reduce the world to a calculable nexus of forces. This does not seem very far from the view of science in analytical philosophy. Heidegger, it is true, gives visuals metaphysical priority and deplores technologically oriented science for its alleged deleterious effect on human flourishing. Analytical philosophers, for their part, tend to avoid such metaphysical and existential questions. We concur that these questions ought generally to be bracketed. But not entirely. Our extended analysis later in this chapter of brain localization visuals illustrates the Heideggerian paradox that the further the life sciences progress toward a Galilean ideal, the further away they move from the understanding of life in Heidegger's sense: "scientific research is not the only manner of Being which this entity [*Dasein*] can have, nor is it the one which lies closest" (1962a, 32).

There is another way we depart from Heidegger's philosophy. We believe that the heart of scientific communication is scientific argument, a subject on which Heidegger is silent, and about which analytical philosophy has much to say. It is to this topic to which we turn in our next chapter.

In this first part of the present chapter, we attempt to locate the "Heideggerian transformation" in astronomy, physics, chemistry, geology, and the life sciences. We do not claim that we have unearthed the exact time of transition; we doubt whether there was an exact time. We claim only that we have unearthed in the case of some prominent scientists early and clear examples of this historic shift in *Weltanschauung*. In the second part of this chapter, we ask whether this shift is evolutionary in the strict sense. To do so, we explore the development of imaging in the journal *Brain* since its inception in the late nineteenth century. In these various representations of the same strategic research site—namely, the human brain—we see a development analogous, but not parallel, to that seen by William Wimsatt and James Griesemer (1989) in Weismann diagrams over a similar period, a development that they claim constitutes evolution in a strict sense.

HEIDEGGER'S FIRST MODERN SCIENCES: PHYSICS AND ASTRONOMY

Physics and astronomy are Heidegger's exemplary sciences. During the seventeenth century, they became "modern," studies according to a new worldview, one in which the world is conceived and grasped as a picture of spatiotemporal magnitudes of motion. Accordingly, Heidegger differentiates the science of Greece and of the Middle Ages from that of the modern age by distinguishing between the occasional theoretical-mathematical depiction of the world, as in Theodoric's fourteenth-century discourse and diagram on the rainbow or in medieval optics generally (Boyer 1987, 118; Lindberg 1978), and the relentless and wholesale objectification of the world as embodied in the theory-infused visuals that meet us now at every turn in the scientific literature. If our interpretation of Heidegger is correct, theoretical-mathematical diagrams are not only central to science as a means of "possible objectification" of the world (1938, 147)—through the precise calculation within an object domain, a procedure all can grasp and all understand as binding on all—but also to the spirit of the modern age.

It might seem that the ancient Greek astronomy of Ptolemy speaks loudly against our interpretation of Heidegger. But this conclusion would be overhasty. Although the many figures in Ptolemy's *Almagest* most definitely express space-time relationships, the fourth category in our taxonomy from chapter 2, they are not intended as pictures of the real world, nor are they evidence that Ptolemy conceived and grasped the world as a picture of spatial-temporal magnitudes of motion. Instead, Ptolemy created a geometrically based model able to account for astronomical observations with reasonable accuracy. His geometric constructions of epicycles and deferents are a means of calculating the positions of the moon, the stars, the sun, and the planets rather than representing their actual motions through space: "it was the Greek ideal, reaching its highest point in the writings of Ptolemy, to construct a model that would enable the astronomer to predict the observations, or—to use the Greek expression—'to save the appearances'" (Cohen 1985, 29). In a passage from his prefatory letter to Copernicus's *On the Revolutions* (1543, vol. 2, xvi), Osiander speaks of Ptolemy when he says,

> Perhaps there is someone who is so ignorant of geometry and optics that he regards the epicycle [calculated smaller orbit superimposed on a larger orbit] of Venus as probable, or thinks that it is the reason why Venus sometimes precedes and sometimes follows the sun by forty degrees and even more. Is there any one who is not aware that from this assumption it necessarily follows that the diameter of the planet at perigee [its nearest distance from the earth] should appear more than four times, and the body of the planet more than sixteen times, as great as the apogee [furthest distance]?

In contrast, in *On the Revolutions* Copernicus plainly means to paint a realistic picture of the universe. Yet his firm commitment to uniform

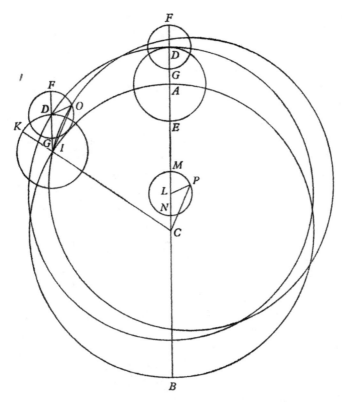

Figure 3.1. Spatiotemporal map of inequality of the solar apsides (Copernicus 1543, vol. 2, 163).

circular motion led to pictures of planetary orbits at least as complex and logically problematic as in the existing Ptolemaic models (Cohen 1985 45–50). Our example is Copernicus's solution to the problem of the inequality of the solar apsides, the sun's positions relative to the earth, positions that vary seasonally. To solve that problem, Copernicus created in figure 3.1 a spatiotemporal map of the orbits of the earth around the sun. He offers two solutions. For each the circle *FG* represents the orbit of the earth, and *C*, the sun. Of his diagram, he says, "Since so many arrangements lead to the same result, I would not readily say which one is real, except that the perpetual agreement of the computations and phenomena compels the belief that it is one of them" (164).

As Edward Rosen makes clear in his translation of *On the Revolutions*, "Copernicus has no grounds for choosing among kinematically equivalent arrangements" (Copernicus 1543, vol. 2, 397). Noel Swerdlow and Otto

Neugebauer (1984, 161) second this: "as he knows very well, he must remain confronted with a variety of models, and can only choose between them *pro posse nostro* [within his own power]." But an astronomy that cannot lead to unique solutions cannot be a science in Heidegger's sense, or in ours. Eduard Dijksterhuis (1961, 68) makes this difference between mathematical and physical astronomy crystal clear:

> the astronomer is free to devise mathematical systems of motion whose results are in agreement with observed facts; the physicist, however, will have to decide whether there is any among these systems which represents what really happens in the heavens; the mere fact of agreement between theory and observation is no guarantee that this is the case.

It is Kepler who will cut the Gordian knot, showing that this Euclidian tangle of epicycles, epicyclets, and deferents can be eliminated only if another conic section substitutes for the circle—the ellipse.

Our next astronomical example, a truly Heideggerian one, also comes from *On the Revolutions*, an exception to the rule. Copernicus explains the apparent retrograde motion of Jupiter; that is, when tracked over the course of a year, the planet seemingly stops, reverses direction, stops again, and proceeds forward. But while the solution is Copernican (vol. 2, 302–4), the illustration we reproduce comes not from Copernicus's great work, but from Galileo's *Two World Systems* (fig. 3.2). Although Copernicus's and Galileo's solutions are geometrically equivalent, it is Galileo who achieves full visual transparency: an image that perfectly embodies the point he wants to make. According to Galileo, this diagram "ought to be enough to gain assent for the rest of the [Copernican] doctrine from anyone who is neither stubborn nor unteachable" (1632, 342). In Galileo's spatiotemporal diagram, the outer arc is a component of the circle of the zodiac in the stellar sphere: imagine it as a screen on which the image of the planet Jupiter is projected. The inner circle is the earth's orbit; the outer, Jupiter's; point O is the sun. The illusion of retrograde motion is caused by the differential orbital circuits of the Earth and Jupiter—the Earth, one year; Jupiter, twelve. When we observe Jupiter from B on Earth, we see it in the circle of the zodiac at position P (follow the line from B on Earth through B on Jupiter to P on outer arc). When we scan to positions C, D, E, and F (corresponding to Q, R, S, and T on outer arc), Jupiter seems to slow down (intervals spanning Q–R, R–S, and S–T become shorter). At earthly positions G and H (corresponding to U and X on outer arc), Jupiter appears to move backward. As Jupiter and Earth continue in their orbits, Jupiter appears to turn around again (Y and Z on the outer arc).

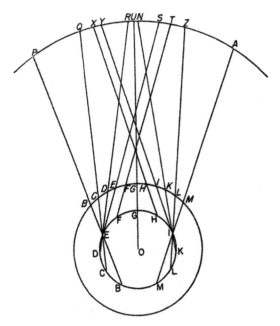

Figure 3.2. Spatiotemporal diagram illustrating retrograde motion of Jupiter (Galileo 1632, 343).

This geometric figure thus makes visible a highly plausible explanation of retrograde motion from a physics standpoint, one that is also consistent with astronomical observations.

Galileo may serve equally well as an example of the Heideggerian transformation of physics. Reproduced in figure 3.3 is a diagram from *Two New Sciences* (Galileo 1638) expressing space-time interrelationships. It is an elegant visual embodiment of a law regarding uniform accelerated motion for a freely falling body. Interpretation of this spatiotemporal representation of a falling body requires considerable work, matching and scanning between the image and text; it requires both perceptual and cognitive effort.

The law is readily stated. Of a body beginning at rest and moving from C to D with uniform accelerated motion, the law states that the time of traversal is equal to the time in which the same space would be traversed by the same body in a uniform unaccelerated motion if its speed were one-half the final speed of the latter body.

The proof is geometrical, so bear with us. Let's begin by reconstructing the geometric figure to the left of line CD. Start with line AB, which represents the time it takes for the uniformly accelerating body to fall from point

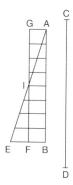

Figure 3.3. Geometric proof of law of uniform accelerated motion (Galileo 1638, 165).

C to D. Next look at line EB. It represents the maximum and final speed of that falling body. Now we can draw a diagonal line connecting E and A, and a series of equally spaced parallel lines above EB within the triangle AEB. Those parallel line segments within AEB reflect the uniformly increasing velocity of the object as it accelerates from A to E. But we are not quite done. We locate the halfway point along line EB and label it F. Then, we construct the rectangle ABFG, where the rectangle segment GF intersects the diagonal AE at I.

We are now ready for the actual proof from the geometric figure. We can infer that rectangle ABFG is equal in area to triangle AEB since triangle IEF fits precisely into the space occupied by GIA. Now, one must imagine that triangle IEF has been rotated and pasted over GIA in such a way that all of the parallel lines in rectangle ABGF reach line GF. Those equal parallel lines within ABFG now represent the unaccelerated motion (uniform velocity) of the body over time AB. And since the triangles IEF and GIA are identical, line segments EF = GA = FB. Therefore, line segment FB is one-half line segment EB. But by definition line segment EB is the final speed of the accelerating body, while line segment FB is the uniform speed of an unaccelerated body.[4]

Once we have followed Galileo's proof through "QED" (165–66), we can see his diagram with fresh eyes. We can visualize a body moving from C to D with uniform acceleration from rest to a final speed (line segment EB), then another body falling through the same space in the same time with a single speed, half the final speed of the first body (line segment FB).

In this diagram, neither the size of the body, nor the length of lines

---

4. See Drake (1972) for a fuller explanation of Galileo's diagram.

CD and AB, nor the number of parallel lines drawn within the triangle and rectangle is Galileo's concern; all that matters is the orientation of the lines to each other, governed by the principles of Euclidian geometry. Why is EB the maximum and final speed? Because we stipulate that this is the case. How many equally spaced parallel lines can we draw above EB? An infinite number. What is important is "Galileo's choice of a physically real and measureable velocity—one-half the terminal speed—in place of a theoretical construct, [such as] the instantaneous middle speed" (Drake 1972, 37). Once you accept the analogy between kinematics and geometry provisionally, you find that the geometrical result magically coincides with the measurable behavior of falling bodies. On this basis, your commitment need no longer be provisional. The diagram is an example of the extent to which Galileo saw geometry not only as a means of solving problems in the real world, but also as an accurate representation of the real structures of that world. That this is indeed his view is clear from a well-known passage from *The Assayer* (1623):

> Philosophy is written in that vast book which stands forever open to our eyes, I mean the universe; but it cannot be read until we have learnt the language and become familiar with the characters in which it is written. It is written in mathematical language, and the letters are triangles, circles and other geometrical figures, without which means it is humanly impossible to comprehend a single word. (Quoted in Crombie 1959, vol. 2, 142)

Consequently Galileo's QED differs from that of Ptolemy; his proof is not a theorem that accounts for the phenomena and is without realist pretensions; it is rather a theorem that explains the behavior of bodies, their actual and potential motions, a step in the uncovering of the causal structure of the world. Setting down such pictures, conceiving and grasping the observable world as a picture "in a mathematical language," is not Galileo's occasional, but his habitual practice—as it was for Galileo's contemporary, Kepler, and for generations of natural philosophers to follow: Descartes, Huygens, Newton, Leibniz, to name just the more illustrious.

While Galileo represents the birth of mathematical physics, Robert Boyle represents its experimental counterpart. It is not that Galileo did not experiment; it is that he did not build elaborate apparatus designed to constrain and manipulate nature. Boyle did; his air pump (fig. 3.4), depicted in his *New Experiments Physico-Mechanicall, Touching the Spring of the Air, and Its Effects: Made, for the most part, in a New Pneumatical Engine*

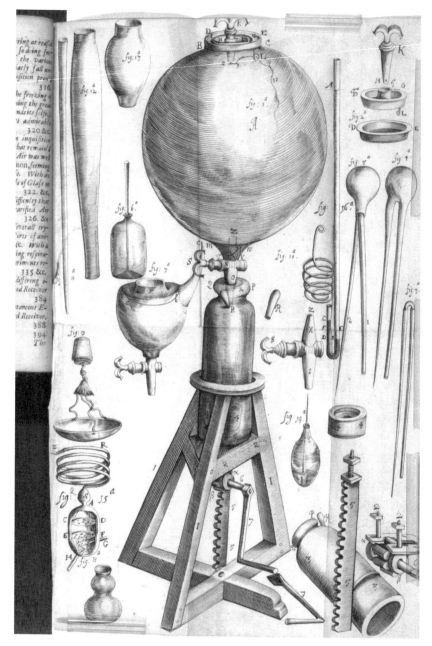

Figure 3.4. Drawing of research equipment: air pump (Boyle 1660, frontispiece).

(1660) and representing the seventh category in our taxonomy, is the remote ancestor of the cloud chamber whose experimental history is the subject of the second chapter of Peter Galison's *Image and Logic: A Material Culture of Microphysics* (1997, 65–141). Whereas the cloud chamber is a machine for discovering facts about the microworld, Boyle's air pump is a machine for discovering new facts about the earth's atmosphere (Shapin and Schaffer 1985, 26–79).

Occupying the privileged center position in figure 3.4 is its three-dimensional representation. It has two main components: a large glass bulb (about thirty "wine quarts" in volume) on top of a pumping apparatus supported by a wooden frame. The very top of the bulb has a port for inserting objects of study (burning candles, different species of animal, Torricelli apparatus). Below the bulb is a hollow brass cylinder containing a piston, which was worked by the crank at the bottom and sucked air out of the bulb or forced more air into the bulb. Around the apparatus at the engraving's center are various parts of the whole along with other devices used in conducting experiments. Boyle's engraving is not intended to stand for all air pumps; it

is not a schematized line drawing [as is Faraday's drawing in chapter 2] but an attempt at detailed naturalistic representation complete with the conventions of shading and cut-away sections of the parts. This is not a picture of the "idea" of an air-pump, but of a particular existing air-pump. (Shapin and Schaffer 1985, 61)

Because of Boyle's experimental successes, however, this image did come to symbolize a robust machine for new fact production in the study of the "spring of the air." Most famously, in Boyle's experiment No. 16, measurements with this machine led to Boyle's law: given a constant volume and temperature, the pressure of air (or "the spring") is proportional to its density.[5]

EMERGENCE OF HEIDEGGERIAN
VISUALIZATION IN GEOLOGY

Antoine Lavoisier's contribution to the transformation of chemistry in the late eighteenth century is well known; less well known is his 1789 contri-

---

5. Boyle does not state this law in the first edition, but in an addendum to the second edition (West 2005, 22, 37).

bution to geology. "General Observations on the Horizontal Beds that Have Been Deposited by the Sea and the Consequences that Can Be Drawn from Their Dispositions Concerning the History of the Earth" is the fruit of over two decades of fact-gathering and reflection. In his article on visualization in early-modern geology, Rudwick concludes that "in the early nineteenth century a new group of visual products became common; and their degree of formalization enables them to be vehicles for what I have termed the structural cognitive goals of the self-styled science of 'geology.' Finally, from about 1820 onwards, these 'structural' forms of illustration began to be developed still further in theoretical directions that enabled them to become a visual language for causal cognitive goals" (1976, 179; his emphasis). This is the goal that Lavoisier seems to have accomplished as early as 1788. In so doing, he exemplifies the vanguard in the transformation of geology into a Heideggerian science that centers on spatiotemporal magnitudes, that is, potentially quantifiable changes in the earth's geology over vast tracts of time.[6]

At the beginning of this memoir, Lavoisier (1789a) defines two new geological terms, "pelagian" and "littoral":

These initial thoughts lead us naturally to the conclusion that there must exist in the mineral realm two sorts of very distinct strata, the one formed in the open sea, and at a great depth, and that I will call, following the lead of M. Rouelle, pelagian strata, the other formed at the coast, and that I will call littoral strata; that these two sorts of strata have distinctive characteristics, which make it impossible for us to confuse the one with the other; that the former invariably present us with a heap of limestone, of the debris of animals, of shells, of marine bodies accumulated slowly and peaceably over a very long succession of years and centuries; that the latter, in contrast, everywhere present us with the image of movement, of destruction, and of tumult. (355)

The remainder of the memoir deals with several research questions about these two types of strata:

---

6. Lavoisier's memoir had little impact on the science of those tumultuous times in France: it was published the same year as the French Revolution erupted. Moreover, Lavoisier's theory ultimately proved "wrong in every detail," in the somewhat hyperbolic words of Stephen Jay Gould (1998), though Lavoisier's many insights into geologic stratigraphy proved far ahead of the times and were not fully acknowledged until the twentieth century (Carozzi 1965). Not all strong arguments for a new knowledge claim stand the test of time, then as now.

How are such diametrically opposed observations to be accounted for? How can effects so different be assigned to the same cause? How can it be that the movement which has worn down quartz, rock crystal, the hardest of stones, which has smoothed their jagged edges, has nevertheless left the most fragile and delicate of shells undisturbed? (352)

In figure 3.5, by superimposing a geometric on a realistic visual code, Lavoisier infers the law-like character of the explanation he favors, the rise and fall of the sea level over time:

the curvature of the sea-floor, from the coast to the open sea, resembles a segment of a parabola, whose axis is parallel to the horizon, that is to say, the inclination of the coast to the horizon, at its juncture with the open sea, is nearly 45 degrees, an inclination that then diminishes up to the point where the sea is absolutely still, a point where its surface tends to become absolutely horizontal. (357)

By imprinting mathematics on nature, figure 3.5 accomplishes a Heideggerian transformation. At the upper center of the figure, Lavoisier superimposes a geometric diagram, a move that permits us to read the realistic depiction below as an expression of changes in spatial relationships over geological time. This superimposition is Galilean insofar as it links geometry to natural phenomena; it departs from the Galilean in that there is no implication that these relationships can actually be expressed in mathematical terms; there is no equation to be solved. In effect, the geometrical scheme permits us to see through the iconicity of Lavoisier's representations to their indexicality, to see their underlying law-like causal structure as a calculable nexus of forces:

First of all it is evident that if the sea encroaches on the coast, if its level rises of an amount equal to BS, the cliff that existed at AB will be undermined at the level of S. Landslides will be frequent until a new cliff HR forms at the SR end of the high-tide line. If sea level continues to rise gradually in amounts equal to ST, TV and VX, the cliff will retreat successively to IQ, KP and so on until the sea will have reached its upper limit. Supposing this limit to be MY, the corresponding cliff would be at LM and the upper surface of the chalk originally along the line of L-K-I-H-A will now be along M-P-Q-R-B. Furthermore, this surface will be covered with beds of pebbles, different sorts of sand, and marls, in

Formation des Bancs Littoraux
dans la supposition ou le niveau de la Mer s'eleve par un mouvement succesif très lent

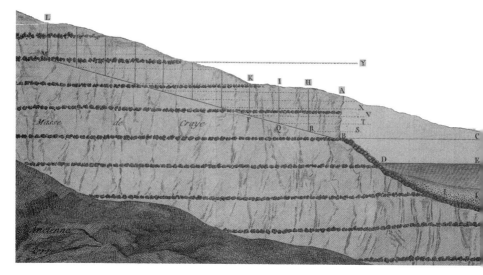

Figure 3.5. Map of geologic site near Paris built out of the hills and valleys of the Seine. The cliff is a "mass of chalk" resulting in a mass of "primary rock" (Lavoisier 1789a, plate 2).

other words, with littoral beds consisting of matter that originated from the destruction of the rocks originally located inside the line L-K-I-H-A-M-P-Q-R-B. In its encroachment on the coast, therefore, the sea is both destructive and constructive. (359–60)

This Heideggerian shift is also evident in Charles Lyell, who in his final work employs geology in the service of the new science of anthropology: among other things, *The Geological Evidences of the Antiquity of Man* (1863) explores the anthropological implications of the geophysical relationships between England and the Continent. Lyell locates the emergence of hominids at the end of an extended period in which what is now the British Isles oscillated above and below sea level—now an archipelago, now a string of islands, now part of the Continental land mass. The geological and biological evidence for this oscillation is abundant. In Scotland, the evidence for submergence consists of "the joint evidence of marine shells, erratics [large boulders randomly scattered and out of place with the landscape], glacial striae [marks of glaciers having passed] and stratified drift at great heights"; in North Wales and central England, of "marine shells"; in Cumberland,

Yorkshire, and Ireland, of glacial striae and erratics; in central England, of marine shells (1863, 274). The biological evidence for the union of the Continent and the British Isles consists of the selective distribution of flora and fauna between the two: "The naturalist would have been entitled to assume the former union, within this post-pliocene period [from 1.8 million years ago], of all the British Isles with each other and with the continent, as expressed in the map [our fig. 3.6], even if there had been no geological facts in favour of such a junction. For in no other way would he be able to account for the identity of the flora and fauna found throughout these lands" (1863, 277).

Figure 3.6 consists of two maps, one superimposed on the other. In the

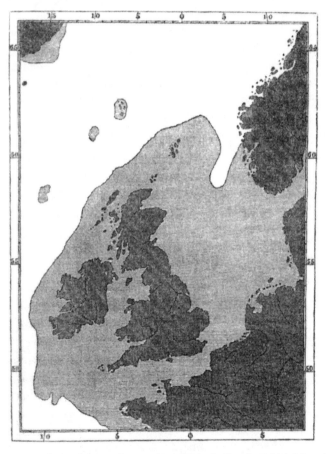

Figure 3.6. Map of part of the northwest of Europe, including the British Isles, showing the extent of sea which would become land if there were a general rise of the area to the extent of 600 feet (Lyell 1863, 279). The darker shade expresses what is now land; the lighter shade, an area of the seafloor just 600 feet below sea level.

background, an extended European continent hovers, surrounding and swallowing the familiar configuration of the British Isles. The foregrounded map is just a map; the map in the background, implied by the evidence Lyell has just adduced, is both more and less than a map. It resembles a map in that, by scanning and matching, the reader can generate propositions such as, "During the ice age, the Continental coast engulfed the British Isles and reached as far as 11 degrees longitude to the west and 62 degrees latitude to the north." But in this case formulating such quantitative propositions is secondary to determining the image's meaning. Key instead is the Gestalt principle of *Prägnanz*: those viewers already familiar with the map of the British Isles and Europe instantly recognize these land masses without verbal explanation. Contrast and figure-ground distinguish the past and present, land and sea. Once provided with the verbal context in the figure's caption, the viewer can instantly grasp this picture's meaning. By superimposition of distant past on the present, Lyell has turned two representations of space into one representation of space-time. Lyell's map is iconic, not only of land-masses and surrounding seas, but of a theory: an oscillating subsidence and elevation taking place over geological time and caused by a calculable nexus of forces that affects the distribution of flora and fauna and the emergence of the human species.

EMERGENCE OF HEIDEGGERIAN
VISUALIZATION IN CHEMISTRY

The Heideggerian transformation in chemistry took place between the late eighteenth and the mid-nineteenth centuries. As we turn from the chemistry of Étienne-François Geoffroy to Antoine Lavoisier, we see a shift in focus from empirically derived chemical affinities to relationships that purport to mirror the order of nature. As we turn from Lavoisier to Dalton, we see another shift in focus: from the order of nature to the nature of the spatiotemporal magnitudes that constitute that order.

Geoffroy's chemical table, reproduced in figure 3.7, focuses on empirically derived chemical affinities. Along its uppermost line, we see a variety of chemical reagents, represented by alchemic symbols defined at the table's base. Below each are substances arranged in order of their affinity. The nearest has the greatest affinity; it is a substance that none of the substances below it would be able to displace. Column sixteen is an example: water can displace alcohol, but the reverse is not possible (Crosland 1978, 240–41). About Geoffroy's table, Roberts (1991) makes the essential point that "nowhere [in it] or in the memoir in which it appeared was the question of the

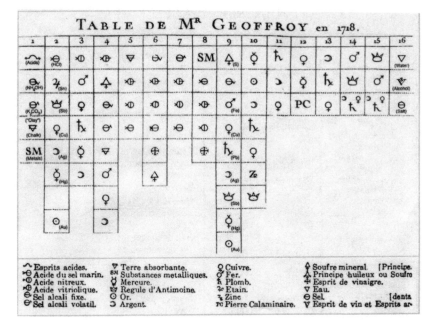

Figure 3.7. Table of affinities among known chemical substances (Geoffroy 1718, 212).

substances' elemental or causal properties broached" (105). Moreover, "the structure of chemistry tables . . . remained essentially the same until the late 1780s" (107).

Though it bears a superficial resemblance to a table, Lavoisier's figure 3.8 is no ordinary table; rather, it is a precursor to Mendeleev's periodic table of 1869, where space stands for relationships among "simple substances" (what we now call "elements"), substances Lavoisier defines as natural bodies that chemical methods can decompose no further. The middle column names all the known simple substances, some of which have "new names" because the old chemistry had given several different names to the same substance, as reflected by the column on the far right. The first column divides all the known simple substances into four general types, depending on observable chemical properties.

Lavoisier's 1789 table systematizes the "simple substances belonging to all the kingdoms of nature," at least as best known by chemical methods at the end of the eighteenth century. Like Mendeleev's, Lavoisier's table is meant to be a true reflection of the elements of which the material world is composed: "the perfection of the nomenclature of chemistry, envisioned in this account, consists in conveying the concepts and the facts truly,

TABLEAU DES SUBSTANCES SIMPLES.

	NOMS NOUVEAUX.	NOMS ANCIENS CORRESPONDANTS.
Substances simples qui appartiennent aux trois règnes, et qu'on peut regarder comme les éléments des corps.	Lumière...........	Lumière.
	Calorique..........	Chaleur. Principe de la chaleur. Fluide igné. Feu. Matière du feu et de la chaleur.
	Oxygène...........	Air déphlogistiqué. Air empiréal. Air vital. Base de l'air vital.
	Azote.............	Gaz phlogistiqué. Mofette. Base de la mofette.
	Hydrogène.........	Gaz inflammable. Base du gaz inflammable.
Substances simples, non métalliques, oxydables et acidifiables.	Soufre.............	Soufre.
	Phosphore.........	Phosphore.
	Carbone...........	Charbon pur.
	Radical muriatique....	Inconnu.
	Radical fluorique....	Inconnu.
	Radical boracique.....	Inconnu.
Substances simples, métalliques, oxydables et acidifiables.	Antimoine.........	Antimoine.
	Argent	Argent.
	Arsenic............	Arsenic.
	Bismuth...........	Bismuth.
	Cobalt............	Cobalt.
	Cuivre............	Cuivre.
	Étain.............	Étain.
	Fer	Fer.
	Manganèse.........	Manganèse.
	Mercure...........	Mercure.
	Molybdène.........	Molybdène.
	Nickel............	Nickel.
	Or................	Or.
	Platine...........	Platine.
	Plomb............	Plomb.
	Tungstène	Tungstène.
	Zinc..............	Zinc.
Substances simples, salifiables, terreuses.	Chaux............	Terre calcaire, chaux.
	Magnésie..........	Magnésie, base de sel d'Epsom.
	Baryte............	Barote, terre pesante.
	Alumine...........	Argile, terre de l'alun, base de l'alun.
	Silice.............	Terre siliceuse, terre vitrifiable.

Figure 3.8. Table of known elements: groupings and names of "simple substances" (Lavoisier 1789b, 135).

omitting nothing, and above all adding nothing; nomenclature must be a faithful mirror, because—and we cannot repeat this too often—it is never nature or the facts that she presents to us, but our own reasoning processes, that deceive us" (1892, 359). Like Mendeleev's, Lavoisier's is not a table for data presentation and retrieval; instead, its purpose is to name and arrange chemical substances in a systematic way and to demonstrate its benefits by contrast with the conventional chemistry of the day. But there is another, equally important aspect of Lavoisier's table, and the many other similar ones in his *Treatise*, ordering various combinations of elements: its structure is meant to be a faithful mirror, not only of the substances themselves, but of the structures existing among them. Even such unknowns as the "Muriatic radical" ("Radical muriatique . . . Inconnu") have their place in Lavoisier's scheme of things, which is also *the* scheme of things, purportedly, the causal structure of the world as meant by chemistry. Lavoisier's table represents the chemical elements as a nexus of forces, calculable in principle and embodied, eventually, in the modern chemical equation. This is not to say that his particular model of nature's working is the final word: "Chemistry progresses toward its goal, and toward its perfection," Lavoisier says, "by dividing, subdividing, and subdividing once again, and we do not know where it will all end. We cannot be sure that what we regard as simple today will be simple in the final analysis" (1789b, 137).

Though obviously of great historical importance, Lavoisier's tables did not become models of modern chemical visualization. At the dawn of the nineteenth century, however, visualization practices in chemistry did undergo a fundamental shift in a Heideggerian direction. For the first time chemists attempted to depict elements and their spatial arrangement within molecules, "even though direct sensual or instrumental access to that level was not available" (Rocke 2010, 2). Lavoisier's calculable nexus of forces was now to be supplemented. The cutting-edge chemists throughout the nineteenth century spent much time and effort determining the identity and quantity of the constituents of various molecules, creating molecular formulas and equations on the basis of their analytical data and observations, and picturing the molecular formulas. By creating these pictures along with an elaborate set of rules governing chemical combinations, these scientists could also predict the existence of previously unknown compounds formed by adding or subtracting elements, which they subsequently sought to synthesize (Rocke 2010, 201). In Heideggerian terms, these chemists were picturing the microworld in the manner of Galileo's astronomy and physics, and Lavoisier's and Lyell's geology.

John Dalton's *New System of Chemical Philosophy* (1808) is an early example of this shift. In figure 3.9, Dalton represented twenty "elements," only fourteen of which survived later scrutiny. They are depicted as circles, each with unique identifying marks to easily distinguish one from the other—putting to good use the Gestalt principles of similarity and contrast. Dalton's rows of elements are not all the elements postulated at that time, but those for which he had calculated the atomic weight, the basis for his scheme of organization. Dalton ordered his elements mathematically by relative weight only, with the lightest element (hydrogen) assigned a weight of 1, and the heaviest element (mercury), 167. According to Rocke (2005), "in order to deduce an element's atomic weight Dalton needed two things: an assumed molecular formula for one of the element's common compounds, and analytical data on that compound." For example, Dalton posited—incorrectly—that water consists of one atom of hydrogen and one of oxygen, which he represented visually by a circle with a dot in the center next to an empty circle (item 21 in fig. 3.9).[7]

Dalton's visual is composed of thirty-seven geometric symbols—each standing for a different element or common compound: for example, water, alcohol, and nitric acid. Dalton arranged the elements in order of increasing atomic weight and the compounds in order of increasing number of elements. In his two-page-long explanatory caption to the figure, Dalton listed the identities of each of the thirty-seven symbolic representations and their relative weights. In the main body of the text (Dalton 1808, 213), he also explained the reasoning behind the pictured combinations, transforming chemistry into combinatorial mathematics:

> If there are two bodies, A and B, which are disposed to combine, the following is the order in which the combinations may take place, beginning with the most simple: namely,
>   1 atom of A + 1 atom of B = 1 atom ["molecule" in modern terms] of C, binary.
>   1 atom of A + 2 atoms of B = 1 atom of D, ternary.
>   2 atoms of A + 1 atom of B = 1 atom of E, ternary.
>   1 atom of A + 3 atoms of B = 1 atom of F, quaternary.
>   3 atoms of A and 1 atom of B = 1 atom of G, quaternary.

7. Note that numbers above the elements in figure 3.9 are image numbers, not atomic weights. The latter are listed in the figure caption, where Dalton reports the atomic weight of oxygen as seven (later revised to eight). Better analytical data soon after Dalton's book was published indicated the molecular formula for water as H_2O, doubling the relative weight for oxygen to the present value of sixteen (Rocke 2010, 5).

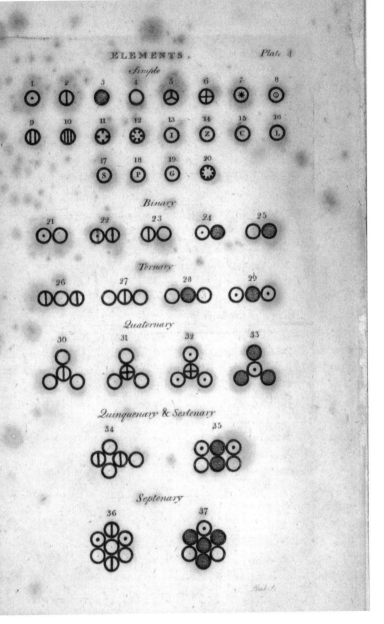

Figure 3.9. Visual representations of elements and compounds (Dalton 1808, facing 219).

Dalton's choice of representation did not catch on. Yet, inspired by his example, later chemists invented alternative visual styles of representation for elements and molecules, among them networks of differently marked circles linked by lines, elemental abbreviations (H, C, Na, Li, etc.) linked by lines, and sausage shapes encasing groups of circles. Throughout the nineteenth century, these images populated chemistry textbooks and research articles. In the first half of that century, chemists could claim no more than that their depictions represented the chemical affinities among the elemental constituents. Only later could they begin to construct convincing arguments that their molecular images depicted a shape corresponding to a structure bestowed by nature (Rocke 2010, 171). In other words, in Heideggerian fashion, the space of chemical visuals gradually evolved; it was slowly transformed from a metaphor for affinities among chemical constituents to a structural map of the microworld.

EMERGENCE OF HEIDEGGERIAN VISUALIZATION IN THE LIFE SCIENCES

We start our investigation here with a typical seventeenth-century visual in the life sciences, a first step toward the realization of a Heideggerian vision of science. We shall see that, while it turns a living being into a biological machine, it does *not* represent that being as a consequence of a calculable nexus of forces. For that transformation to begin in a serious way, we must wait for Darwin. Figure 3.10 is a two-panel realistic drawing of a chameleon and some of its body parts. In the top panel we see what the scientists saw after dissecting the animal depicted in the bottom panel. As Perrault explains, the chameleon "is shown alive, perched on a tree a little tilted towards the side which it mounts, so as to reveal as much as is possible of the top of the head and underside of the stomach" (1669, 16). The whole animal dissembled into its numbered parts transforms an object of nature into an object of science.

The chameleon is the most salient element in the bottom drawing: no labeling is necessary to make that clear. The dominant Gestalt principle is *Prägnanz*; readers immediately recognize an animal posed on a tree. It occupies the privileged position of the center, perched on a tree before a hilly region of what appears to be France. The background's only purpose is to foreground the animal. It has no scientific meaning; it does not represent the animal's prototypical habitat, the tropical. Scanning to the semiotic components immediately above the realistic depiction, we see a skeleton in a similar pose; above that, we see various other internal organs labeled with

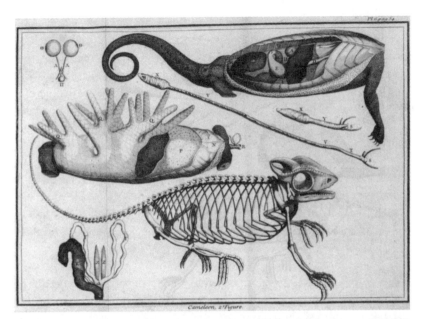

Cameleon, 2.e Figure.

Cameleon, 1.re Figure.

Figure 3.10. Realistic drawing of animal: Perrault's chameleon (1669).

letters. For full comprehension of the upper panel, we must scan and match, moving between the key on the previous page and the labeled bodily parts in the upper plate. We identify the chameleon's eyeballs on the upper left, a cross section of the guts in the upper right, tongue below that, and so forth.

The text following the figure is more than an extended verbal description of the visual displays. In it, Perrault compares the physical attributes of his chameleon to reports by other naturalists, arguing that in many cases they were mistaken:

> Pliny is mistaken when he makes it as large as a crocodile . . . The head of our specimen was an inch and ten lines [twelve lines to an inch]: from the head until the commencement of the tail, it was four and a half inches; the tail was five inches; and the feet were each two and a half inches long.

In this context, this illustration becomes something more than an artistic rendering; it becomes a direct challenge to other natural history texts, one based on disciplined observation.

The eighteenth-century work of Carl Linnaeus (1735) represents another step in the direction of a life science that realizes Heidegger's vision. But his systematic work, based on the sexual characters of his plants, is a mathematization analogous to that of Ptolemy in astronomy. There is no sense that such a classification can be regarded as biologically fundamental. In contrast, in Darwin's *Origin of Species* (1859) this concept is fundamental; the one visual in his book embodies it. From this visual, properly interpreted, we can infer that Perrault's chameleon is the product of evolutionary forces: "in biological evolution, individuals among species interact with one another and the environment. When they do, in sexual species at least, the genetic endowments of individuals combine to produce a next generation, a new cohort on which natural selection has operated and continues to operate in the direction of more nearly satisfactory adaptation to each other and the environment" (Gross, Harmon, and Reidy 2002, 216). When this does not happen, when they do not adapt, individuals die and species become extinct. The calculable nexus of forces that creates, destroys.

In figure 3.11 lines and curves depict aspects of the world only insofar as these embody evolutionary forces. In our classification, the figure is a flowchart or process diagram in which space is twice a metaphor: in its vertical dimension it represents time; in its horizontal dimension, organic diversity. It is crucial that the diagram has gone halfway to meet theory, divesting itself in the process of those features that facilitate pattern recognition in the

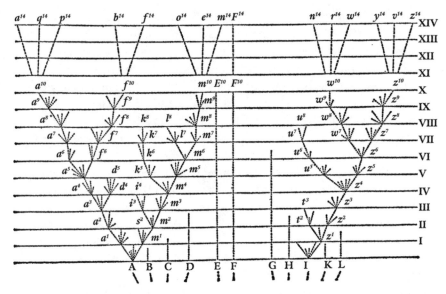

Figure 3.11. Diagram of evolutionary theory showing divergence of taxa (Darwin 1859, end papers).

case of middle-sized objects such as real trees and real geological columns. It is precisely this divestment that permits the diagram to mediate between the world and its explanation, to represent the world and the forces acting on it.

The figure's central message is that organic diversity increases over time. The diagram is interpreted by scanning up from A, B, C . . . , and matching with time intervals I, II, III . . . , under the guidance of the accompanying text. The further up viewers go, the nearer they get to the present; the further apart species are, the more they have differentiated themselves from their parent species through evolutionary processes. Reading the diagram vertically from bottom to top, we trace descent. We see that a^1 and m^1, and a^{10} and m^{10} have the same remote ancestor A, where the superscript indicates the era corresponding to its vertical position. Reading horizontally, we see that far greater evolutionary divergence exists between the latter pair than between the former.

Extinction, the ultimate price that evolutionary forces exact, is also represented in the diagram, literally, by the end of a line. Species s^2, the descendent of A, becomes extinct in era II; species u^8, a descendant of z^4, which is, in turn, a descendant of I, becomes extinct in era VIII.[8] The dia-

8. The break between eras X and XI is unexplained. We assume it was meant to indicate the verbal equivalent of "and so on and so on" indefinitely.

gram also embodies the cause of extinction, the struggle for existence. It is no accident that A, F, and I, whose ancestors are still living, were originally nearly the most divergent from each other, for the nearer species are along the horizontal dimension, the more likely they are to compete for the same territories and resources in the battle for survival, and the more likely they are to become extinct. In many cases—B, C, and H, for instance—there is no evolutionary change over time, but only in the case of F do we still have a living species, a living fossil. In other cases, the branches of evolutionary trees eventually crowd out these lines of descent.

In the text, Darwin makes a theoretical point about the significance of the numbered letters: "In our diagram the line of succession is broken at regular intervals by small numbered letters marking the successive forms that have become sufficiently distinct to be recorded as varieties. But these breaks are imaginary, and might have been inserted anywhere, after intervals long enough to have allowed the accumulation of a considerable amount of divergent variation" (119). Despite their individuation as points, species a^1, a^2, a^3, etc., are not real; even the lines of divergence are conjectural; they are *dotted* lines. Only the processes that create divergence are real. Evolution embodies an ontology of process, rather than product.

Moreover, as the time scale of the process is open, the amount of evolutionary change each incremental coordinate represents may legitimately vary. Darwin says, "If we suppose the amount of change between each horizontal line in our diagram to be excessively small, these three forms may still be only well-marked varieties . . . but we have only to suppose the steps in the process of modification to be more numerous or great in amount, to convert these three forms (a^{10}, f^{10}, and m^{10}) into well-defined species" (120). But the time scale can also change enormously. Darwin asks us to imagine the vertical increment represents "a million or hundred million generations" (124). This variation in time scale and, consequently, in the span of evolutionary change blurs distinctions between species and variety, genus and species. In effect, Darwin turns the diagram into an argument for these intellectual leaps.

Thus far we have seen Darwin's diagram as a representation of evolution; but it is equally a representation of the geological record, "a succession of the successive strata of the earth's crust including extinct remains" (124), a record of the evolutionary forces of the past. In the first case, the vertical dimension is metaphoric: space represents time. In the second case, the vertical dimension is denotative: space represents only space—the geological column. On this reading, the diagram is an example of the third category in our taxonomy. In another reading, however, the diagram exemplifies the fifth category: space is a metaphor for temporal sequence. The geological column is also, as it were, time made visible; its vertical dimension also represents time. Really, it *is* time: looking at the fossil embeddings in the geological column, we are, in a very real sense, looking at the products that past forces created. It is a past, moreover, that the first reading of the diagram allows us to understand. Looking at the geological record, Darwin says, "we find . . . such evidence of the slow and scarcely sensible mutation of specific forms, as we have a just right to expect to find" (336). The location of these forms and their fossil remains is not their physical but their evolutionary location, not where they are found, but where, once found, they belong in the evolutionary sequence.

We can see the argumentative value of the visual presentation when we compare Darwin's diagram with its verbal "equivalent"; Darwin ends his fourth chapter with an extended simile based on the analogy between evolutionary development and the growth of a tree. Up to a point, the comparison holds:

> The green and budding twigs may represent existing species; and those produced during each former year may represent the long succession of extinct species. At each period of growth all the growing twigs have tried to branch out on all sides, and to overtop and kill the surrounding twigs and branches, in the same manner as species and groups of species have tried to overmaster other species in the great battle of life. (129)

But the argument from analogy soon outwears its theoretical welcome. In the verbal expansion, disanalogy undermines Darwin's theoretical purpose: In the case of the real tree that must be the vehicle of Darwin's extended simile, the lower branches, trunk, and roots must be alive or there can be no top branches. The minimalism of his diagram permits Darwin to evade this fatal implication. The diagram represents only those characteristics of a tree that serve his theoretical purpose: representing the process and products of the potentially calculable nexus of evolutionary forces. Darwin's diagram is Heideggerian because, although not yet mathematized in the strict sense, it is open to calculation: "This designation does not mean that the principles themselves are mathematical belonging to mathematics, but that they concern the mathematical character of natural bodies, the metaphysical principles which lay the ground of that character" (Heidegger 1935–36, 148–49; 1962b, 190–91). Such later formulations as the Lotka-Volterra predator-prey equations (Berryman 1992) testify to the fulfillment of this potential.

We may say with some justice that, so interpreted, Darwin's diagram *is* his theory, his argument. On this view, once we understand the diagram, we understand Darwin. On this construal, our view of the *Origin* as a book with one diagram is mistaken; it is really a diagram whose caption happens to be as long as a book. Such a view, however, misconstrues the division of labor that the *Origin* so elegantly exemplifies. Necessarily, Darwin's diagram falls short of depicting the complexity of the processes involved. It altogether omits sexual selection and the effects of inbreeding that would retard change: "The diagram illustrates how natural selection, the principle of divergence, and the evidence for the extinction of many species in time [can be conceived and grasped as a single picture of a calculable nexus of forces]. It goes far in exemplifying the complexity Darwin needs, but cannot do it all. Darwin needs the rest of the book, not [only] to explain the diagram, but to explain to his readers the complexity of the process that is missing from the diagram."[9]

For the reasons discussed above, Darwin's diagram is not fully Heideggerian. While it represents nature as a calculable nexus of forces, it does not represent its constituent features as spatiotemporal magnitudes. For this to occur in the life sciences in general, we have to await the developments in the twentieth century.

DO VISUALS EVOLVE? THE EXAMPLE
OF BRAIN LOCALIZATION

We have located a point of "Heideggerian transformation" in astronomy, physics, chemistry, geology, and the life sciences; we have not raised the question of whether the images in these sciences evolved in the strict sense. To do so, we explore the development of imaging in brain localization research as manifested in select images from *Brain*, a journal founded in 1878 specifically to advance the locationist program correlating specific regions of the brain with effects on other parts of the body (Star 1989, 32; Young 1970, 198). Our concern will not be the conceptual development of this research program in neuroscience and the importance of instruments of detection in this enterprise; that will be assumed. Our concern will be, rather, the exemplification of this program as the final step in a Heideggerian transformation of the life sciences, realized visually in a shift from realistic drawings and photographs to brain scans and line graphs that betoken existence in a world in which a calculable nexus of forces operates on spatiotemporal

9. Michael Reidy, personal communication, 2002.

magnitudes. Coincident with this shift, we also see the realization of a Heideggerian paradox: the further the life sciences progress toward a Galilean ideal, the further away they move from understanding of what it is to be a living thing. As we move through the decades of *Brain*, from its founding to the present, we see the enactment of this paradox.

We begin in 1835, the year Franz Joseph Gall announced a research program that would link brain activity to significant biological and psychological functions:

> Whoever would not remain in complete ignorance of the resources which cause him to act; whoever would seize, at a single philosophical glance, the nature of man and animals, and their relations to external objects; whoever would establish, on the intellectual and moral functions, a solid doctrine of mental diseases, of the general and governing influence of the brain in the states of health and disease, should know that it is indispensable that the study of the organization of the brain should march side by side with that of its functions. (Gall cited in Young 1970, 247–48)[10]

Gall did not follow through on this program; neither did David Ferrier, one of the founders of *Brain*, a journal expressly devoted to localizing biological and psychological phenomena in the brain, uncovering in each case the causal nexus of behavior. Despite this slow start, contemporary authors publishing in *Brain* have attained heights of scientific rigor unimaginable to Gall's contemporaries. Researchers have moved from probing the dead to probing the living; from a scalpel wielded in autopsy to *f*MRI (functional magnetic resonance imaging) that maps a living brain, a far more sensitive and noninvasive probe; from individual observation to controlled experiment; from tentative localization hypothesis to localization theory. It is these achievements that we sample in the form of articles from *Brain*.

In the earliest of the articles in our sample (Magnan 1878–79), we see

---

10. Young's quotation is from François Joseph Gall and J. C. Spurzheim, *On the Functions of the Brain and Each of Its Parts: with Observations of the Possibility of Determining the Instincts, Propensities, and Talents, or the Moral and Intellectual Dispositions of Men and Animals, by the Configuration of the Brain and Head*, vol. 2 (Boston: Marsh, Capen, and Lyon, 1835), 45–46. This is a translation of *Sur les functions du cerveau et sur chacune de ses parties* (Paris: Ballière, 1822–25). It was published under the auspices of the Boston Phrenological Society. Gall's first name is listed as François rather than Franz, his given name, because the publication is in French; apparently, it is not a translation of a German original.

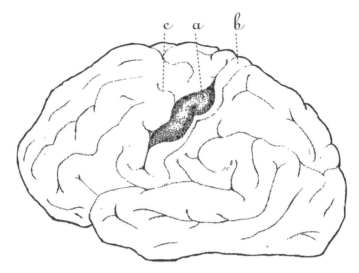

Figure 3.12. Drawing of human brain with cerebral tumor: (a) angiolithic sarcoma, (b) atrophy of the upper two-thirds of the ascending parietal convolution, and (c) ascending frontal convolution (Magnan 1878–79).

a lithograph of the human brain we reproduce as figure 3.12. It depicts a tumor, labeled *a*, in the brain's left hemisphere, a growth rendered salient as a dark shape against a pale ground. While the depiction is clearly iconic, its function in creating scientific meaning is indexical. It is as a cause of behavioral impairments that it makes its appearance. Unfortunately for the proponents of localization, the subject's impairments were also on the left side of her body, the side opposite the one the tumor should have affected. To Magnan, this anomaly is easily explained as "a benign, indolent tumor, developing very slowly" (565), too easily explained, perhaps, since localization in any case was bound to fail as a hypothesis: the effects observed— seizures and intellectual deterioration—are too heterogeneous and general to make such localization of a causal nexus plausible.

Whereas Magnan represents the brain by means of lithography, Bolton (1903), writing a quarter century later, does so by photography, as exemplified by the photograph reproduced as figure 3.13.[11] With photography, evidential value is enhanced: we see not an artist's rendition, but an actual

11. Photography was available to Magnan, but not used by him. As a technology, it may have been at a stage too primitive to capture the image he wanted. There also may have been problems with reproducing a photograph in a journal.

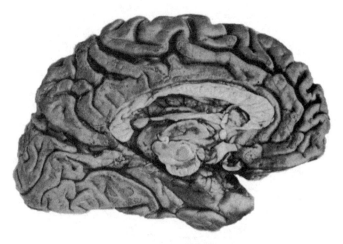

Figure 3.13. Photograph of the inner surface of the left hemisphere (Bolton 1903, 228).

adult brain—presumably what our eyes would see at an autopsy, mediated only by the technology of the camera.[12] Nevertheless, the real advance in evidential value comes not from photographic realism, but from scientific realism: an accompanying diagram (fig. 3.14), organized conceptually, maps the functional areas of the brain, its causal nexuses, potentially calculable. It is this diagram that tells us *how* to see the photograph; it tells us *what* is salient.

Only when we superimpose this map of brain function on Bolton's photograph, as instructed (228), do we see the dramatic results that autopsy reveals: "the anterior center of association is both greatly wasted and acutely changed. The temporal and precuneal centres of association are acutely changed. The visual projection centre is intact" (228). Moreover, "the amount of cerebral wasting and the associated morbid changes inside the cranium . . . *vary directly with the amount of dementia existing*" (237; our emphasis). As a consequence of mapping spatial relationships within the brain, then, the photograph of the brain becomes for the scientist a photograph of the mind. At the same time, it becomes the index of such

12. We understand that this view of photography is naïve, but we presume that the author of the article and his readers shared this naivete. For a nuanced analysis of the changing perception of "objectivity" with the invention of machines like the camera for automatically transforming nature into knowledge, see Galison's "Judgment against Objectivity" (1998) and "Objectivity Is Romantic" (1999).

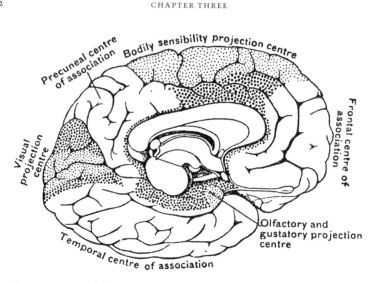

Figure 3.14. Map of the inner surface of the left hemisphere (Bolton 1903, 228).

heterogeneous behaviors of the "insane" as "absence of self-consciousness (as regards others)," "extreme vanity," "callous and selfish behavior," and "delusional states" (238).

There is a problem, however, with the causal sequence implied. In Bolton's map, Gestalts of enclosure and contrast give an unwarranted impression of firm borders. Moreover, as a consequence of the superimposition of the map onto the photograph, this precision is transferred to a causal nexus: the relation of these areas—and their impairment—to the heterogeneous behaviors Bolton characterizes as dementia. This same difficulty is exhibited a quarter century later in an autopsy of a patient who had hemichorea—random involuntary jerky motions in the distal parts of the limbs on one side of the body (Weil 1928; figure not reproduced). But this particular medical researcher is willing to err only on the side of caution. After analyzing his case and placing it in the context of others reported in the literature, Weil feels that "conclusions which state that the lesion which causes hemichorea must lie in a specific part of [the brain] are, at present at least, premature" (45).

Writing a quarter century later, Humphrey and Zangwill (1952) exhibit a similar caution. Like Magnan and Weil, they focus on an individual and produce a case study. But there is also a distinct departure from previous studies in the sample: Humphrey and Zangwill's focus is on a living subject. Furthermore, that subject—an officer suffering from a brain injury sustained on the front lines in World War II—has himself become a scientific instrument, cooperating with the researchers in the investigation of the causes of his own disability. He is asked to draw a map of England from memory, reproduced as figure 3.15.

The most salient feature of the map shown in figure 3.15 is England's west coast. Its *left* side is entirely absent, a startling violation of the Gestalts of enclosure and continuity and a clear indication to the researchers of an impairment caused by a *right*-sided occipito-parietal brain injury. From the patient's point of view, the drawing is iconic—he is drawing a map of England. From the point of view of the researchers, however, it is indexical. This indexicality is confirmed by brain activity data recorded by EEG (electroencephalogram), a noninvasive technique that "showed a gross focal anomaly in the right posterior parietal and posterior temporal regions" (Humphrey and Zangwill 1952, 316). In effect, the patient's pencil has functioned like the recording pen of a scientific instrument: the map of England is a map of his damaged brain. We see in this article a clear shift from the qualitative to the data-driven, a shift that opens the door to theory and to Heideggerian transformation; in effect, the map of England and the EEG have made visible as a calculable nexus of forces, hitherto invisible,

Figure 3.15. Map of England drawn from memory by psychiatric patient (Humphrey and Zangwill 1952, 319). Reprinted with permission from Oxford University Press.

spatiotemporal magnitudes of motion, traces from which theory may eventually be inferred. Over time, the man's behavioral deficit persisted, despite a general recovery sufficient to resume normal work. This pattern, Humphrey and Zangwill assert, has implications for a general hypothesis of cerebral dominance; it "[suggests] that the shift in dominance for symbolic functions [from the left] to the right hemisphere did not convert the [right hemisphere] into the major hemisphere for visual recognition and topographical skill" (313).

By 1978, Ongerboer and Kuypers can attain an even greater degree of precision than Humphrey and Zangwill can, a further shift in the data-driven direction. They focus on a curious physiological response first noted in 1896: a tap on one side of the forehead induces "reflex blinking." This blinking effect has two typical components: an early response related only to the side of the brain that is stimulated, and a late response related to both sides. Ongerboer and Kuypers studied the latter response. Their subjects were thirteen patients with Wallenberg's syndrome, difficulty in speaking or swallowing caused by the occlusion of the posterior inferior cerebellar artery or one of its branches supplying the lower portion of the brain stem. In type-C cases—the most severe—"stimulation of the nerve on the affected side failed to elicit a bilateral late reflex response and stimulation of the nerve on the unaffected side only elicited a late response on that side but not on the affected side" (287).

Figure 3.16 depicts these data trends in two pairs of brain scans, made visually clear by means of the Gestalt principles of continuity, similarity, and contrast. The labels R (right side of brain) and L (left side) group each pair of scans. In these graphs the first arrows in both sequences indicate the stimulus, while the second arrow in the first sequence indicates the late response; "mV" indicates the amplitude of the response in millivolts on the y-axis; "ms," the time of the response in milliseconds on the x-axis. Although the graph records brain events in their temporal order, the organization of the visual as a whole is conceptual: it counts as evidence of neuropathology. In the first pair, nerve stimulation on the right side produces a late blink response on that same side. By contrast in the second pair, stimulation on the left side produces no late response on either side. The graphs themselves are indexical; they give us a precise indication of the cause of the brain abnormality, an infarction depicted in the photograph reproduced as figure 3.17, an area of tissue death made visually clear by means of the similarity of the two sides as a whole and by means of the contrasting area within the left side.

Although they depict the same thing, the graph and photograph differ in

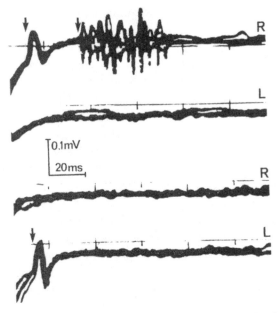

Figure 3.16. Graphs indicating type C abnormality of the late blink reflex on left side (Ongerboer and Kuypers 1978, 289). Reprinted with permission from Oxford University Press.

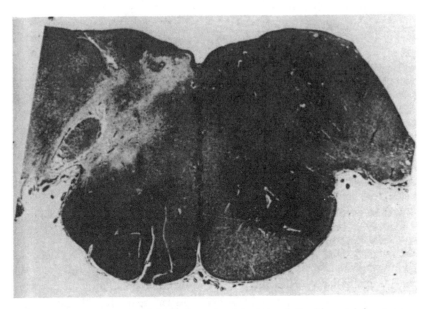

Figure 3.17. Cross-section through the patient's brain (medulla oblongata) showing a unilateral infarction on left side (Ongerboer and Kuypers 1978, 288). Reprinted with permission from Oxford University Press.

relationship to their object of study. The photograph's relationship is direct: we see a freeze-frame of what the researchers *saw*. The graphs' relationship is indirect: we see what the researchers *detected* by measurement of brain activity. In terms of science, however, each has equal ontological status; in fact, this equal footing is the point of the science that depicts them as cause and effect. The photograph is iconic of a brain lesion—indeed, its nature as evidence of causality requires this iconicity. It shows an infarction that "involves mainly the spinal fifth nerve complex" (289), an exact localization of the cause of the abnormality represented in the graphs. These are indexical in two senses. First, they point to a cause; second, their shape is consistent with that cause, electrical brain activity: the effect of the infarction is mirrored precisely in the lack of effect that the bottom pair of graphs depict. Most salient is the difference between these and their upper counterparts, a salience that is, equally, visual and conceptual. Furthermore, the photograph and its accompanying graphs form a sequence with diagnostic value: "the late blink reflex has value, because the results may provide the clinician with further information about the extent of the brain-stem lesion" (292). When we compare figures 3.16 and 3.17 with figures 3.13 and 3.14, we can see how far brain localization research has come in seventy-five years in picturing the brain and its operation as a calculable nexus of forces.

Figure 3.18, a Heideggerian apotheosis, comes from an article on impaired visual processing by James and four colleagues (2003). Central to the research process that creates this suite of images are two factors: (1) a theory that the cerebral cortex has two discrete visual processing pathways or "streams," ventral and dorsal, the source of a particular calculable nexus of forces, and (2) a powerful and noninvasive method, functional magnetic resonance imaging (fMRI), which measures minute changes in blood flow, in spatiotemporal magnitudes, caused by neural activity.

The subject of this investigation is a woman, D.F., who, as a result of a brain injury, is incapable of discriminating among even the simplest geometric forms, although she has a normal capacity when reaching for and grasping objects. In effect, she feels, but cannot see their shapes. On the basis of behavioral evidence alone, Goodale and Milner had conjectured "that the form-processing network of D.F.'s ventral system ha[d] been all but destroyed" (James et al. 2003, 2471), while the functions of her dorsal stream remained largely intact. James et al. submit this hypothesis to a rigorous test, an experimental design comparing D.F.'s performance with that of normal subjects. If this hypothesis were confirmed, James and his coauthors aver, it would provide support for the view that discriminating and grasping are "served by separate cortical systems" (2470). Furthermore,

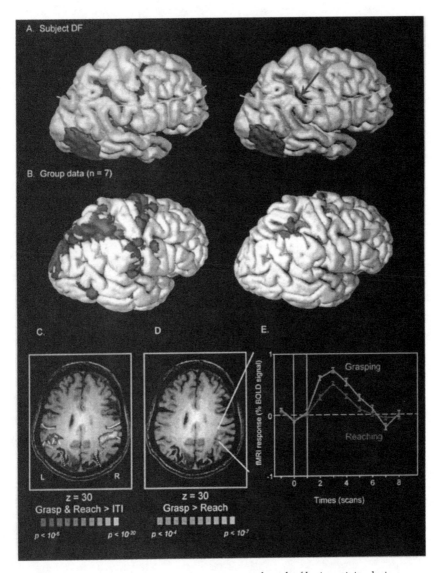

Figure 3.18. Magnetic resonance image maps and graph of brain activity during grasping and reaching (James et al. 2003, 2473). Original in color (shaded regions in brain maps on left are red; those on right, green). Reprinted with permission from Oxford University Press.

it would display "the powerful role fMRI can play in revealing the pattern of spared and compromised functions in neurological patients" (2470). Figure 3.18 dramatizes the fact that the dorsal visual processing system has indeed remained intact in D.F., a dramatization made particularly striking by the use of color, making good cognitive use of the Gestalt principle of contrast.[13] In the original color figure in all of the brain maps, red patches indicate activation due to D.F.'s grasping and reaching motion; and green, activation due to grasping versus reaching motion.[14]

In figure 3.18, Gestalt enclosure differentiates each image visually from every other. Each is displayed as on a museum wall. At the same time, the framing of the whole presents to viewers a single visual unit with a single conceptual message: "D.F., like seven neurologically intact control participants, showed clear activation in the dorsal-stream action areas during reaching and grasping" (2469). Gathered in one figure are a computer reconstruction of the brain, a "surgical slice" through the brain, and a graphic representation of D.F.'s reaching and grasping response—different views of the same object. Nevertheless, because each of its components is on the same ontological plane, despite these differences, the figure as a whole embodies the causal nexus of the normal brain activity involved in reaching and grasping. In its interactive entirety, the figure supports the claim that D.F.'s grasping and reaching are wholly normal: her ventral stream is seriously impaired, leading to her inability to discriminate geometric objects, but her dorsal stream is functionally intact.[15] And being generated by computer algorithms and analytical machines, the image itself is the product of a calculable nexus of forces.

Within a scientific context, three problems arise from images such as this. The first concerns the epistemic reach of the technology employed; the second, the extent to which individual variation affects the interpretation of the data; the third, the integration of the isolated brain functions into the full spectrum of human behavior. Research programs, such as those of James et al., rely heavily on MRI technology, a dependency that leaves their

---

13. While current technology still does not permit us to reproduce this figure inexpensively in color, the color version is readily available in academic libraries and on the web. Fully to understand the points we are making, it must be consulted.

14. "Grasping requires both transport of the hand to the target and pre-shaping the hand and fingers to reflect the visual properties of the object, such as shape, size, and orientation, whereas reaching requires only transport of the hand" (Cavina-Pratesi, Goodale, and Culham 2007).

15. The more extensive "colored" areas in the brain image representing the control group data (image B) are a statistical artifact.

mappings open to the charge that they confuse precision with accuracy. No one questions the rough correlation between behavior and MRI readings, but some point out that the blood flow we actually measure is far slower than the neuronal activity that we wish to measure (Dobbs 2005). These critics feel that the precision that MRI researchers exhibit does not match the accuracy they can legitimately achieve with the technology they routinely employ. The wrong spatiotemporal magnitudes are being measured.

Individual variation in brain localization is a second problem: it undermines any general conclusions based on an individual case, such as that of D.F. The language area, for example, shows considerable individual variability:

> What has not been appreciated previously is the marked individual variability in the location of the mosaics essential to language. . . . This variability is quite marked in all areas except the most posterior portion of the inferior frontal gyrus. Even there, enough variability is present so that the classical Broca's area is occasionally not involved in language. This variability is probably the explanation for the difficulty in determining the exact location of the Wernicke language area from the extent of the temporoparietal lesions producing aphasia. . . . The Wernicke language area of the classical model is clearly an artifact of these essential areas in different patients, for rarely if ever are essential language areas found in an individual patient. Indeed, the entire extent of the classical Broca and Wernicke language areas is seldom essential for language in an individual patient. (Ojemann et al. 1989, 324)

There is a third problem that arises in a scientific context. In the end, researchers measure only what they can measure: isolated brain functions and their behavioral correlates. But no one believes that the significance of brain activity lies only in the identification of these isolated functions. Although careful researchers, such as James et al., avoid assigning the higher-level functions that constitute our humanity to specific areas of the brain—avoid creating a new phrenology—their program holds little promise of giving us insight into that humanity.

There is a last problem, one whose context is broader than that of science. Whatever the differences between the brain and the person whose brain it is, surely the most salient is that persons are the products not only of their brain activity, however integrated, but also of an environmental complex that includes their life experiences and the intersubjectivity of their particular life world. Ignoring this philosophical commonplace, neuroscientific researchers tend to "share the *Idea* that the brain must *be* in some fundamental way the person" (Dumit 2004, 103; his emphasis). The fact is that their methods preclude their understanding of human beings as persons. Theirs is a version of Heidegger's paradox: to see human beings as mechanisms is to treat them neither as human nor as beings.

The bias of brain researchers is explained in part by the extraordinary quality of the images they routinely produce. In the century and a quarter since the founding of *Brain*, the means of investigating brain function and the means by which this function has been visualized have co-evolved with remarkable synergy. As an unintended consequence, the causal arrow has pointed in two directions: from theoretical and methodological development to visual embodiment, and from visual embodiment to theoretical and methodological development. It is the second direction that has been epistemically problematic. Speaking to historian Joseph Dumit, scientist Michael Ter-Pogossian says, "you select images that prove your case. However, the case is also proven, *supposedly*, in your text" (Dumit 2004, 96; our emphasis). Dumit extends this comment to support Roland Barthes's thesis that "the image no longer illustrates the words; it is the words which, structurally, are parasitic on the image": PET scans (and of course *f*MRI scans) "participate in [a] reversal of veridictory authority" (Barthes cited in Dumit 2004, 143). The excess of precision over accuracy simply overwhelms when it is given visual embodiments as striking as those of James et al.

DO VISUALS EVOLVE? THE EXAMPLE
OF WEISMANN DIAGRAMS

Our survey of the visual history of brain localization raises a question: Is evolution as we employ it in this chapter's title anything more than a metaphor for development? Can a case be made that the development of scientific visuals is conceptual evolution in the strict sense, that is, differential survival in the face of selection pressures that a theory of evolution imposes, a changing conceptual environment that shapes a changing population of visual species? To Griesemer and Wimsatt (1989), at least, scientific diagrams are visual species that can be viewed as units of selection in conceptual evolution. As such, they constitute adaptations, cost-effective means for increasing the probability of survival of the concepts they embody. They increase this probability by reducing "demands on memory, computation, or other limited resources" (99). Griesemer and Wimsatt's example is the sequence of diagrams that represents August Weismann's germ theory of

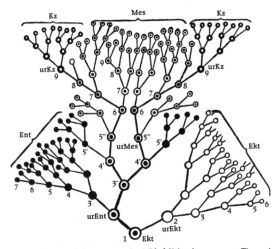

Diagram of the germ-track of Rhabditis nigrovenosa. — The various generations of cells are indicated by Arabic numbers, the cells of the germ-track are connected by thick lines, and the chief kinds of cells are distinguished by various markings: — the cells of the germ-track by black nuclei, those of the mesoblast (Mes) by a dot in each, those of the ectoderm (Ekt) are white, those of the endoderm (Ent) black; in the primitive germ-cells (ur Kz) the nuclei are white. The cells are only indicated up to the twelfth generation.

Figure 3.19. Diagram of germ-track of *Rhabditis nigrovenosa* (Weismann 1893, 196).

heredity. In figure 3.19, Weismann's founding tree diagram of 1893 concerns the lineage of the germ-plasm within the context of the individual development: it depicts a single organism and a single generation. Like Darwin, Weismann employs space as a metaphor for evolution, where time flows from the bottom to the top. Readers can start at any cell and follow any one of the germ tracks upward in time as it develops.

Weismann's diagram has had many progeny. One of several on which Griesemer and Wimsatt comment extensively is John Maynard Smith's representation of heredity in his *Theory of Evolution*, reproduced as figure 3.20. In Maynard Smith's diagram, while space is still a metaphor for cell creation, time flows from left to right. While the diagram's visual components are still geometric, they now represent not organic, but theoretical entities. It is the lower section of this diagram that reveals the real purpose of Maynard Smith's visual adaptation; in it, the conceptual parallel is made explicit between the continuity of the germ-plasm (G) as a source of phenotypes (P) and the continuity of DNA as a source of proteins. Maynard Smith's is an adaptation of Weismann's diagram with the specific purpose of creating a historical and conceptual link between theories he regards as analogous.

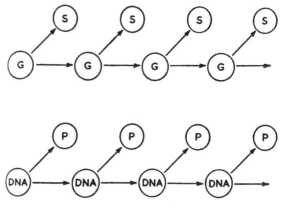

Weismann and the central dogma

Weismann and the central dogma. Reprinted from *The theory of evolution* by J.
Meynard Smith (2nd ed., 1965; Figure 8, p. 67) with the permission of Penguin Books Ltd.,
London, copyright (©) John Meynard Smith 1958, 1966, 1975.

Figure 3.20. Diagram of J. Maynard Smith's view of the central dogma of molecular
genetics (Smith 1965, 67).

Simultaneously, Maynard Smith reaches back to the pioneering work of
Weismann and moves forward to the cutting edge of then-current molecular
biology.

Griesemer and Wimsatt conclude that the diagrams they have examined
"show a number of interesting patterns, including descent without modification, descent with modification, differential proliferation, adaptive radiation, extinction, relict survival in a changed and specialized niche, and successive simplification for efficient specialization to a simplified niche" (128).
Maynard Smith's diagram, an example of the second of these categories, exemplifies that "what is passed on [in visual succession] is a relatively compact generative structure which, in the right conceptual and social environment, generates the theory-phenotype of the next generation" (89).

We do not think that the visuals of brain localization that we have just
examined meet this high evolutionary standard; in no case do our visuals
form an evolutionary lineage in a strict sense. Nonetheless, we would not
dismiss a claim to evolutionary status solely on that account. We would
contend that in the case of brain localization differential survival in the face
of selection pressures is present, though it operates not on a specific lineage,
as with Griesemer and Wimsatt, but on *Weltanschauungen*, worldviews in
which the sciences of any era must find their place.

In this chapter, in astronomy, physics, geology, chemistry, and the life sciences, we have illustrated the emergence of the era of the scientific world picture, a worldview in which a calculable nexus of forces operates on spatiotemporal magnitudes. In the case of brain localization, we have illustrated the development of this worldview within a specialized area of research, a development that Griesemer and Wimstatt also illustrate with Weismann diagrams and their progeny. These two philosophers of science demonstrate that strict evolution among scientific visuals is possible; we show that an analogous form of evolution may be general. In our next chapter, we explore the arguments that secure scientific claims in the era of the world picture whose evolution we have just traced.

CHAPTER FOUR

Verbal-Visual Interaction and Scientific Argument: The Contexts of Discovery and Justification

But in science the credit goes to the man who convinces the world, not to the man to whom the idea first occurs.
—Frances Darwin, *Eugenics Review* (1914)

. . . if I had to choose between your wave mechanics and the matrix mechanics [of Heisenberg-Born], I would give preference to the former, owing to its greater visualizability [*Anschaulichkeit*].
—H. A. Lorentz to E. Schrödinger (1926)

The arguments we see in the books and articles that constitute the primary scientific literature are the end points in a process of discovery and justification, a chain of reasoning that leads from hypotheses about the way the world might be to arguments about the way the world is. We offer a model for the pursuit of scientific truth in these two contexts. At each stage of the model of this process we favor, vigilant problem-solving manifests itself: a rational decision must be made, one for which, if asked, scientists can give reasons regarded as plausible by their fellow researchers. At stage one, the decision is to attack a particular problem; at stage two, to pursue a particular hypothesis; at stage three, to solicit acceptance of a particular claim on the basis of the arguments set out in its favor. From the point of view of the generating scientists, the process is entirely teleological: persuasive argument is always the end point.

We are aware that the distinction between the contexts of discovery and justification is fraught with controversy (Schickore and Steinle 2006). Many philosophers of science deny its existence; and most, its utility. Whatever the philosophical status of the two contexts, their argumentative status, the only status that interests us, is clear. The context of discovery is a private arena, shared within research teams and among invisible colleges; scientists know they will not be held accountable for the views they share in these conversations and emails. In contrast, the context of justification is a public arena in which authors take full responsibility for the soundness of their arguments and the truth of their claims. While discovering is private, discovery requires the acknowledgment of peers. There is no such thing as a private discovery (Gross 1998).

In the arguments that constitute the primary scientific literature, we see two very different semiotic components in play: the verbal and the visual. The verbal includes both words and numbers, which can also evoke mental imagery; the visual consists of actual depictions or their symbolic counterparts. That a division of semiotic labor between the verbal and the visual occurs can be true even in those articles in which actual depictions are, apparently, absent: Röntgen's first paper (1895) announcing his discovery of X-rays contained no physical images; his discovery, of course, was exclusively a matter of images. There are also articles that contain no images—for example, Einstein's first four papers on Brownian motion. But such theoretical papers are most perspicuously construed as a component of a suite that encompasses their confirmation: Jean Perrin's confirmation of Einstein's theory of Brownian motion (1909) is image-laden. The ubiquity of images in scientific texts does not mean that science can never proceed in the absence of the visual, but it does mean that no model of scientific argument can be adequate that does not incorporate both words and images as potential components at every stage of inquiry.

We also contend that scientific argument transforms. As the objects and events of inquiry pass through the stages of the model we advocate, there are alterations in semiotic valence: these objects and events shift from the iconic to the symbolic and indexical, from representing and generalizing about the world to explaining how the world works. By means of argument, Watson and Crick turn X-ray diffraction photographs of their object of study, DNA, into a plausible stereochemical realization of its structure, a shift within the iconic mode. But this shift is also a transformation from the iconic to the symbolic. Because the Watson-Crick model applies to any DNA molecule, each of its realizations stands for the rest. There is a further transformation: the molecule's structure suggests a scheme of replication, a shift in mode from the iconic to the indexical, from description to explanation. Thus scientific argument is Heideggerian in two senses: it En-frames a picture of the world in which a certain kind of rationality is consistently privileged, and it frames discovery and justification as, literally, processes in pursuit of which we must see accurately in order to understand.

THE CONTEXT OF DISCOVERY: PROBLEM SELECTION
AND HYPOTHESIS GENERATION

In *Experience and Prediction*, Hans Reichenbach (1938) introduced a distinction between the context of scientific discovery and the context of its justification, "between the thinker's way of finding [a] theorem and his way of presenting it before a public" (6). Decades later Thomas Nickles (1980) suggested a useful amendment to Reichenbach's distinction. For Nickles, the context of discovery involves not one but two stages: the generation of hypotheses and the decision to pursue one among those generated. We would like to suggest a further amendment. To plunge immediately into hypothesis generation is to begin the story of scientific discovery *in medias res*. The story should begin with the selection of a problem.

Problem selection has as its first step an assessment of the state of knowledge in a field. In general, selection depends on finding a gap, an anomaly, or an inconsistency in the literature (Swales 1990, 141); occasionally, as in Röntgen's discovery of X-rays, selection depends on an unexpected encounter with nature (Friedland and Friedman 1998). The fact that there is a gap, an anomaly, an inconsistency, is a tribute to the rational progress of science up to that point, a progress with which the scientist in question must be fully conversant; the surprise Röntgen experienced depends on the defeat of expectations, an experience impossible without a thorough understanding of the subject in which Röntgen was immersed.

Once a gap, an anomaly, or an inconsistency is detected, once surprise is experienced, scientists must assess whether the human and financial resources they can amass are adequate to its resolution. For Einstein, facing the challenge of the perihelion of Mercury or the paradox inherent in James Clerk Maxwell's equations, it was a matter of one man working alone in the study for an extended period; for Matthew Meselson and Franklin Stahl experimentally determining the means by which DNA reproduces, it involved months of effort working with an expensive piece of equipment, the ultracentrifuge, a scarce resource so in demand that time on it had to be reserved, even on weekends. For all three, it involved a calculated risk, a gamble that their investment of time and resources would pay dividends.

That the choice of a problem itself involves a rational decision makes the literature on social psychology relevant. Janis and Mann (1977) argue for a normative standard in dealing with such decisions—vigilant problemsolving. In science, this method consists of a close interrogation of the state of the field and the problem-solving resources that might be made available. For this method to work well, there is an optimal emotional state: "A moderate degree of stress in response . . . induces vigilant effort to scrutinize the alternative courses of action carefully and to work out a good solution, provided the decision maker expects to find a satisfactory way to resolve a decisional dilemma" (51; emphasis omitted). The absence of any stress whatever signals an absence of urgency, and therefore of motivation; too great a feeling of stress induces hypervigilance, a state whose most extreme manifestation is panic. In all cases, stress is induced by the fear of failure. Scientists are faced with "uncertain risks as well as known costs with regard to money, time, effort, emotional involvement, reputation, morale, or any other resource at the disposal of the decision maker or his organization" (69).

Once a problem is chosen, a hypothesis must be generated which, it is hoped, will lead to its solution. Before we can discuss this stage of our model as rational, we encounter a formidable obstacle. All hypotheses are generated by the same process: induction, the deliberate movement from observations to generalizations and laws. Owing to its centrality to science, there have been numerous attempts to justify induction as a logical process parallel to deduction. But despite the heroic efforts of John Stuart Mill, Charles Peirce, and Norwood Russell Hanson, the justification of induction has proved elusive. David Hume's argument in *Enquiry Concerning Human Understanding* (1748) has survived essentially unscathed. Over two and a half centuries have passed since Hume argued that induction is based not on reason, but on custom and experience:

From a body of like color and consistency with bread, we expect like nourishment and support. But this surely is a step or progress of the mind which wants to be explained. When a man says, *I have formed, in all past instances, such sensible qualities, conjoined with such secret powers,* and when he says, *similar sensible qualities will always be conjoined with similar secret powers,* he is not guilty of a tautology, nor are these propositions in any respect the same. You say that the one proposition is an inference from the other; but you must confess that the inference is not intuitive, neither is it demonstrative. Of what nature is it then? To say it is experimental is begging the question. For all inferences from experience suppose, as their foundation, that the future will resemble the past and that similar powers will be conjoined with similar sensible qualities. If there be any suspicion that the course of nature may change, and that the past may be no guide for the future, all experience becomes useless and can give rise to no inference or conclusion. It is impossible, therefore, that any arguments from experience can prove

this resemblance of the past to the future, since all these arguments are
founded on the supposition of that resemblance. (51; Hume's emphasis)

We may leave Hume's problem, the justification of induction, to philosophers who wish to follow in the footsteps of Mill, Peirce, and Hanson;
in this chapter, we will concern ourselves with argument theory, not with
epistemology, with the justification not of induction, but of particular inductions. We will concern ourselves, in other words, with the challenges
scientists face in generating hypotheses, deciding which to pursue, and integrating what they uncover into arguments designed to survive the scrutiny
of their peers. This alteration in focus has a welcome consequence: it turns
philosophers' attempts to describe and justify induction into a set of clues
to the reasoning that scientists actually employ in their justifications. An
exploration of this basis will show that, although induction cannot be justi-
fied, inductions can. This does not indicate any inroad on Hume's problem.
By enumerative induction, we may move confidently from this and that
white swan to the generalization that all swans are white; by an analogous
process, we may move confidently from particular measurements of the
pressure and volume of a gas to the law that pressure and volume are in-
versely proportional. But no collection of swans or of gas and pressure readings provides sufficient reason for either the generalization or the law.

In this path, we follow Hume. Hume was unconcerned about the practical consequences of his argument: "My practice, you say, refutes my doubts.
But you mistake the purport of my question. As an agent, I am quite satis-
fied in the point; but as a philosopher who has some share of curiosity, I will
not say skepticism, I want to learn the foundation of this inference" (1748,
52). Hume also thought he had discovered just how induction works: "This
transition of thought from the cause to the effect proceeds not from reason.
It derives its origin altogether from custom and experience" (67). And for
the success of our practices, the ways we run our private and professional
lives, this, he felt, was sufficient.

Our exploration of scientific argument, then, will involve consideration
of scientific practices, the ways scientists run their professional lives in the
laboratory and in the field. We will contend that in the contexts of discovery
and justification these practices are exercised according to the dictates of
reason, not in Hume's deliberately narrow sense of deductive implication,
but in a broader sense, one whose criteria are constituted by the collective
judgment of scientific peers. We suggest a social definition of what counts
as rational.

We acknowledge that our view that hypothesis generation is rational in any sense has been authoritatively contravened. Speaking of the formation of fruitful hypotheses, William Whewell says "these are commonly spoken of as felicitous and inexplicable strokes of inventive talent; and such, no doubt, they are. No rules can insure us similar success in new cases; or can enable men who do not possess similar endowments, to make like advances in knowledge" (1984, 211). Carl Hempel's twentieth-century comment on the problem echoes this earlier view: "What determines the soundness of a hypothesis is not the way it is arrived at (it may even have been suggested by a dream or a hallucination) but the way it stands up when tested" (1965, 6). Despite these strong assertions to the contrary, the practice of four centuries of modern science provides us with constraints that keep science's inductive leaps within the bounds of reason as defined by the common sense of particular scientific communities. As Nickles says, these hypotheses are generated in accord with a set of constraints that includes "previous theoretical results, which function as consistency conditions, limit conditions, derivational requirements, etc. There are general methodological demands and perhaps content-specific programmatic demands as well" (1980, 36; emphasis omitted).

To explore these constraints further, it will be helpful to differentiate induction in general from one of its more interesting types, especially relevant to science. The inference that all swans are white is an induction, but so is Kepler's hypothesis that the orbit of Mars is elliptical, and Darwin and Wallace's that a struggle for existence is the linchpin of biological evolution. This seems to us an unfortunate conflation. It seems prudent to give these latter, far more heroic leaps a special name to mark their significant role in the generation of scientific hypotheses. Peirce calls these leaps "abductive." He characterizes their psychological component as follows: "The abductive suggestion comes to us like a flash" (1955, 304). In this process, he says, "the different elements of the hypothesis were in our minds before; but it is the idea of putting together what we never before dreamed of putting together which flashes the new suggestion before our contemplation" (304). In other words, we are surprised, an effect that is always the result of external circumstances: "it is as impossible for a man . . . to give himself a genuine surprise by a simple act of the will" (292).[1]

---

1. It is the sense of sudden revelation that differentiates abduction from the perceptual judgments with which it is otherwise analogous: "I see an inkstand on the table: that is a percept. Moving my head, I get a different percept of the inkstand. It coalesces with the other. What I call the inkstand is a generalized percept, a quasi-inference from percepts, perhaps I might say a composite-photograph of percepts" (Peirce 1955, 308–9).

Abduction is not only a psychological experience; it is also a step in building a scientific argument. So to describe it, however, we must show how scientific practice confines abduction within rational bounds. It does so by means of the methodological and theoretical constraints to which Nickles refers. At the level of method, for example, analogy is a powerful means of justifying one's abductions. Astronomers from Ptolemy on had assumed that the methods of geometry and those of astronomy were analogous. Experience had continually seconded this intuition. However radical, Kepler's solution to the orbit of Mars makes use of this same consensual analogy: the mistaken solution to the Martian orbit, the circle, and the correct one, the ellipse, are both conic sections; indeed, a circle is merely a special case of the ellipse. To us, this means that Kepler's hypothesis of the elliptical orbit of Mars, however innovative in itself and radical its implications, was generated at the routine intersection of geometry and astronomy, the shared social and intellectual space to which Nickles alludes. To say this is not to diminish Kepler's achievement; it is only to contextualize it, to explain why the hypothesis could have occurred to Kepler as an astronomer of his time: the training and practice of astronomers of his time created the possibility of this solution. And of its acceptance.

Analogy also functions at the level of theory. The independent discovery of natural selection by Darwin and Wallace was experienced in each case as a flash of insight. Darwin said that the insight "struck" him (1958 [1887], 42); Wallace, that it "suddenly flashed upon [him]" (1905, 361). But in both cases, the psychological phenomenon was the outer sign of an inner argumentative shift, the inference that an analogy existed between Thomas Malthus's explanation of the fluctuations in human populations and the explanation of fluctuations of populations in the wild. Darwin speaks of "being well prepared" by Malthus "to appreciate the struggle for existence which everywhere goes on from long-continued observation of the habits of animals and plants" (42). Wallace speaks of his discovery that Malthus's "self-acting process" in the wild "would necessarily improve the race because in every generation the inferior would inevitability be killed off and the superior would remain" (362).

While the experience of insight belongs solely to the individual, its possibility lies in the community of which the individual is a member. It is no happy accident that natural selection was discovered independently by these two men. As Robert Merton (1973, 375) has shown, it follows from the institutionalization of science that although "the community of scientists is a dispersed rather than a geographically compact collectivity," nevertheless, it "respond[s] to much the same social and intellectual forces that impinge upon them all." This is why "the pattern of multiple discoveries in science is in principle the dominant pattern rather than the subsidiary one" (356). Even the literary form in which the moment of discovery is routinely cast is a fixed genre, a shared resource as old at least as the narrative of Paul's conversion in the *Acts of the Apostles*.

The social nature of scientific discovery disabuses us of the bias in favor of the individual that leads to the view of Whewell and Hempel, the view that the moment of insight is decoupled from the social and intellectual resources of communities. Speaking of Einstein's discovery of general relativity, Reichenbach (1938), no friend of inductive rationality, makes the vital point—that the resources the scientist employs are community property: "the gift of seeing lines of smooth interpolation within a vast domain of observational facts is a rare gift of fate; let us be glad we have men who are able to perform in respect to the whole domain of knowledge inferences whose structure reappears in modest inferences which the artisan of science applies in his everyday work" (367). The point is not that anybody could have made these discoveries, but that when somebody did, it would be on the basis of shared social and intellectual resources. Absent this shared social and intellectual world, it is difficult to see how such discoveries would ever be accepted as such.

Hypotheses that might be worth pursuing may be realized verbally or visually, and the latter may be described rather than depicted. In his *Notebooks*, Darwin *sees* evolution as "a force like a hundred thousand wedges trying to force . . . every kind of adapted structure into the gaps . . . in the oecomomy of Nature, or rather forming gaps by thrusting out weaker ones. The final cause of all these wedgings, must be to sort out proper structure & adapt it to change" (1987, 375–76). But Darwin also thought in terms of actual images. In his *Notebooks*, he begins a sentence with the words "I think," and ends it not with a noun phrase, but with an image (fig. 4.1). In this diagram, the circled *1* represents an origin; the *A*, *B*, *C*, and *D*, species or genera; and these abbreviations and the image itself work together to represent key evolutionary concepts: descent, diversity, and extinction. Descent is represented linearly, diversity by horizontal distance, extinction by branches without a foot. Through many iterations (Gross 2006, 81–97) as Darwin's thinking evolved, this image ultimately turned into figure 3.11 in chapter 3. Such visual hypotheses as figure 4.1 are "hot cognitions," generated in a flash: "if a person suddenly sees a blazing fire in his fifth-floor apartment, he may visualize himself trapped in the elevator with the power shut off and then see himself, in brief fantasy, emerging safely onto the sidewalk outside the door at the bottom of the staircase" (Janis and Mann 1977, 65).

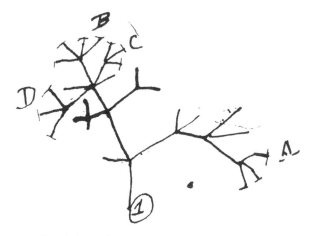

Figure 4.1. Darwin's initial tree diagram of evolution (Darwin 1987a, 180).

The history of science provides us with numerous examples of the value of such hypotheses, a persistent presence consistent with an emerging consensus concerning the epistemic and communicative importance of scientific images. Among historians, we single out Martin Rudwick (2005, 8), who says that the study of visuals ought to be "an ordinary and indispensible part of any attempt to understand scientific work." Among philosophers, we single out Ronald Giere (1996, 272), who "wants to present (part of) a theory of scientific reasoning in which visual presentations of both data and theory can play a significant role." Among sociologists, we single out Michael Lynch (1990, 153), who says that "visual displays are distinctively involved in scientific communication and in the very 'construction' of scientific facts."

THE CONTEXT OF DISCOVERY: HYPOTHESIS PURSUIT

Once a hypothesis is generated, scientists must decide whether its pursuit is worthwhile. To assist in revealing the criteria scientists might employ in exercising this aspect of vigilant problem solving, we will rely heavily on Peter Lipton's insightful *Inference to the Best Explanation* (2004). We shall not enter into a discussion of the merits of Lipton's contribution to the long-standing debate among philosophers concerning the best description and analysis of scientific explanation. We shall leave it to philosophers to weigh its merits against those of Mill's methods, Popper's conjectures and refutations, Hempel's deductive-nomological explanation, or the hypothetico-deductive model. We shall focus only on the insights Lipton gives us into the criteria scientists use in choosing a hypothesis worthy of pursuit.

What does "best" mean in Lipton's title? It means, he says, the "loveliest," as distinct from the likeliest explanation. Lipton's choice of adjective is so unsettling it makes one archaeologist cringe: Lars Fogelin (2007) insists that we talk instead of "compelling" explanations. But Lipton's choice is deliberatively unsettling in the interest of clarity. The loveliest hypothesis is one that exhibits a preponderance of tokens of epistemic promise, which Lipton calls explanatory virtues. Lipton does not draw the parallel between his virtues and the Greek *aretai*, human excellences of any kind. But the parallel seems compelling. Lipton's long list of explanatory virtues consists of elegance, simplicity, predictability, precision, openness to confirmation or disconfirmation, fruitfulness, wide scope, specification of a mechanism, unification of existing facts and theories, fit with existing background beliefs, and variety in the sources of supporting evidence (2004, 66, 122, 168, 173).

Why does loveliness trump likeliness in the choice of a hypothesis to pursue? If an explanation is likely, its persuasive force relies on its capacity to square best with received opinion. This is an explanatory virtue, to be sure; nevertheless, by itself, this concurrence means the pursuit of such a hypothesis is most likely to overlook those more interesting possibilities that received opinion routinely marginalizes or obscures. A likely explanation for the fall of objects heavier than air is that they seek their natural place at the center of the universe, a position coincident with the center of the earth. This explanation is not unlovely: it exhibits simplicity and predictability, as well as fit with existing background beliefs. But Newton's *Principia* offers a lovelier explanation by far. While his theory conspicuously lacks an explanatory virtue, Lipton mentions—it does not specify a mechanism—it nevertheless possesses many others in his list: elegance, precision, predictability, simplicity. An additional virtue of this explanation is its unification of the theoretical insights of Galileo and Kepler. In achieving this unification, this explanation exhibits still another virtue—wide scope: it covers everything from the fall of an apple to the orbit of the moon. It is also fruitful: for example, more than 150 years after Newton's *Principia*, it led to the discovery of Neptune, a discovery made as a result of a mathematical prediction based upon Newton's laws of motion and gravity and an observed perturbation in the orbit of a known planet, Uranus. The explanation is fruitful in another sense: an anomaly in the perihelion of Mercury persisted; Newton's laws could not explain it. Eventually, as science progressed, this anomaly led to an explanation that superseded Newton's in virtue, general relativity. What had been an anomaly in the one explanation became a confirmation of the other. A final virtue of general relativity was its specification of a mechanism.

To the virtues that Lipton mentions, two others may be usefully added. Einstein (1959, 7) provides an insight into the value of the first of these, the potential for visualization:

> What precisely is "thinking"? When, at the reception of sense-impressions, memory-pictures emerge, this is not yet "thinking." And when such pictures form series, each member of which calls forth another, this too is not yet "thinking." When, however, a certain picture turns up in many such series, then—precisely through such return—it becomes an ordering element for such series, in that it connects series which in themselves are unconnected. Such an element becomes an instrument, a concept. I think that the transition from free association or "dreaming" to thinking is characterized by the more or less dominating rôle which the "concept" plays in it. It is by no means necessary that a concept must be connected with a sensorily cognizable and reproducible sign (word).

In an account parallel to Einstein's, Matthew Meselson describes the visual reasoning by means of which he and Franklin Stahl arrived at the experimental design that determined how DNA reproduced. It was a design that relied crucially on visual differentiation among sediments:

> The first experiment we did was at dinner. We had sugar on the table . . . and I just put sugar, a *lot* of sugar, in a glass, and filled it with water and then cut off a piece of fingernail and dropped it in—just to see if you could float, in a solution like that, materials of the density of DNA. I mean, we didn't know exactly the density of DNA . . . but the fingernail seemed a reasonable analogy. And as I remember, the fingernail sinks, even in the strongest sugar solution. [Then we added salt (giving salt and sugar) but the fingernail still sank.] So we needed something denser. We went to [a large oilcloth periodic table that hung] in the guest room and said, "Well, we want something like table salt, sodium chloride, but very dense," so we read straight down the chart from sodium to the heavy elements that are chemically similar—sodium, potassium, rubidium, and then there's cesium, which is the last naturally occurring element in the group. (quoted by Holmes 2001, 175; emphasis his)

We see the drive toward visualization even in cases that would at first glance seem to be unvisualizable, subatomic interactions in light of quantum theory: Feynman diagrams provide an example of the attempt to overcome this supposedly insuperable difficulty. While no one would claim that images are integral to all scientific thinking, it seems incontrovertible that scientific thinking, sometimes of the highest order, occurs in the absence of words.

A second virtue, economy, is also worth adding to Lipton's list. Peirce (1955, 216) makes it clear why:

> Perhaps we might conceive the strength, or urgency, of a hypothesis as measured by the amount of wealth, in time, thought, money, etc., that we ought to have at our disposal, before it would be worth while to take up that hypothesis for examination. In that case, it would be a quantity dependent on many factors. Thus a strong instinctive inclination toward it must be allowed to be a favouring circumstance and a disinclination an unfavourable one. Yet the fact that it would throw a great light upon many things, if it were established, would be in its favour. . . . The expense which the examination of it would involve must be one of the main factors of its urgency.

Lipton omitted this virtue, probably because it is not explanatory in his sense. Nevertheless, to omit it is to ignore an important factor that enters into every stage of hypothesis development, from the decision to pursue to the decision to accept. It concerns not epistemic promise per se, but rather the likelihood that this promise will be realized in a particular case. In other words, it does explain; it is one reason why a scientist may pursue or decline to pursue a hypothesis.

Now we come to a problem Lipton ignores: the relationship of these virtues to one another. Certainly, these virtues may also combine synergistically to enhance the loveliness of any hypothesis. Kepler's hypothesis of the Martian elliptical orbit exhibits wide scope, a single virtue with a single explanatory target: every phenomenon to which it applies is a planetary and lunar orbit. But wide scope may combine with other virtues. The greater the variety of evidence a hypothesis explains, the lovelier it is; additionally, the greater this variety, the greater the hypothesis's potential for eventual unification under a more general explanation. As a result of this process, Kepler's hypothesis becomes a special case of Newton's laws. Virtues can also be at odds with one another: fruitfulness and fit with existing background beliefs exemplify this potential conflict. The more complete this fit, the less likely it is that a hypothesis will be fruitful: the discovery of a new hummingbird or new element exhibits this relatively uninteresting though not unimportant virtue. However satisfying such discoveries are, they will in all probability not be fruitful.

There is another characteristic of the explanatory virtues ignored by Lipton: the presence or absence of degrees. The unconditional endorsement of a hypothesis depends only on its perceived truth, a concept that does not admit of degree. This absence of degree is also a characteristic of other virtues: a mechanism is either specified or not; a hypothesis is either visualizable or not. On the other hand, there can be degrees of elegance, simplicity, fruitfulness, scope, precision, predictability, economy, openness to disconfirmation, variety of supporting evidence, unification of existing theories, and fit with existing background beliefs. Moreover, fit may be either methodological or conceptual, a distinction important because each category contributes independently to a discovery's initial and eventual acceptance. It undoubtedly helped with initial acceptance of Watson and Crick's hypothesis that it did not involve methodological innovation: the X-ray diffraction photographs they used in solving their problem were an accepted means of determining the structure of crystalline forms. And it likely hindered the acceptance of the inferences Galileo made from his telescopic observations that they involved both conceptual and methodological innovation.

The virtue of economy also admits of degrees. All things being equal, it makes sense for scientists to pursue those hypotheses that would be easiest to eliminate if false. But all things are seldom equal. Kepler ignored this rule by investing years of seemingly unrewarding labor in making sense of Brahe's astronomical observations; Darwin ignored it in his pursuit of evolutionary theory over many decades; Watson and Crick ignored it in their pursuit of the structure of DNA despite the disapproval of their supervisor. That each of these tasks may well have ended in failure is a testimony to the temper of investigators more interested in the possibility of gain than the probability of failure.

The explanatory virtues also come in two very different flavors: aesthetic and cognitive. Elegance is an example of an aesthetic virtue. When scientists appreciate a hypothesis because it is elegant, they are savoring it as if it were a fine wine or a work of art; in so doing, they are exhibiting their scientific taste. This taste, however, is not epistemically irrelevant; experience has taught science that beauty is, at times, allied to truth. In his account of the discovery of the DNA double helix, James Watson (1968, 205) makes clear the distinction between the elegance of a hypothesis and its truth:

Lacking the exact X-ray evidence, we were not confident that the configuration chosen was precisely correct. But this did not bother us, for we
wished only to establish that at least one specific two-chain complemen-
tary helix was stereochemically possible. Until this was clear, the objection could be raised that, although our idea was aesthetically elegant, the
shape of the sugar-phosphate backbone might not permit its existence.
Happily, now we knew that this was not true, and so we had lunch, telling each other that a structure this pretty just had to exist.

Precision is an example of a cognitive virtue. Unlike elegance, it is integral
to confirmation or disconfirmation: the more precise a prediction, the more
impressive its confirmation and the more devastating its disconfirmation.
Regardless of whether scientists are appreciating an aesthetic virtue or ad-
miring a cognitive one, they are inferring a connection—a defeasible connection to be sure—between loveliness and truth.

We now move from the characterization of the scientific virtues to the
choice of a hypothesis to pursue. Competing hypotheses may be considered
simultaneously or sequentially. Rachel Laudan (1980) gives us the example
of J. Tuzo Wilson, who, over a period of years, considered as an explanation of the mechanism behind the earth's surface features the rival and in-
compatible hypotheses of contraction, expansion, and continental drift. In
general, however, it is only rival research programs or the same research
programs over time that test rival hypotheses. Prior to the discovery of the
structure of DNA, Pauling had proposed a triple helix using model building;
Wilkins and Franklin analyzed X-ray diffraction photographs of DNA with
the hope of discovering the structure by this means alone; Watson and Crick
preferred a strategy that reconciled the results of model building with those
of the X-ray diffraction photographs. In their work, Watson and Crick did
not consider rival hypotheses at the same time, but one after another. It was
only by dint of the repeated failures of earlier models that they uncovered
their loveliest hypothesis, one that exhibited the virtues of elegance, variety
of supporting evidence, simplicity, unification of otherwise scattered facts,
specification of a mechanism, and fruitfulness. Economy also entered into
the discovery of the structure of DNA: collaboration permitted the winning
team to expand their intellectual resources at the lowest possible cost. It is
no accident that the discovery was made by a biologist and a physicist with
the willing cooperation of a theoretical chemist, Jerry Donohue, and the re-
luctant cooperation of an X-ray crystallographer, Rosalind Franklin.

Theirs was also a hypothesis that lent itself to elegant visualization, exemplified by the bonding structure sketched in a letter Watson sent to Max

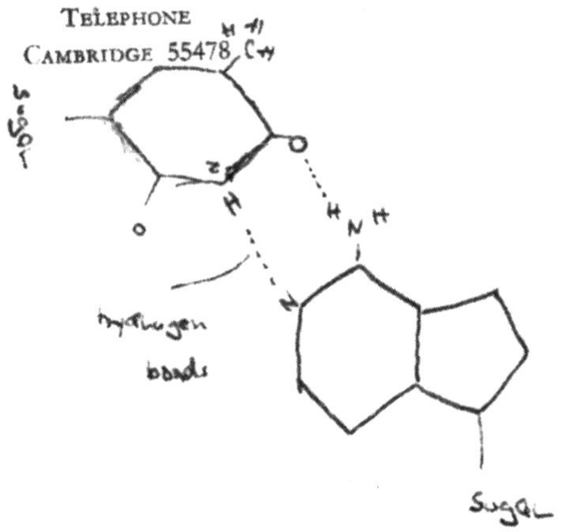

Figure 4.2. Watson's sketch illustrating the spatial arrangement in the bonding of the
thymine-adenine base pair (Watson 1968, end papers).

Delbrück (fig. 4.2) and verbally conveyed by Watson's account, from *The
Double Helix* (1968, 196; our emphasis):

> I suspected that we now had the answer to the riddle of why the number
> of the purine residues exactly equaled the number of pyridimine resi-
> dues. Two irregular sequences of bases could be regularly packed in the
> center of a helix if a purine always hydrogen-bonded to a pyridimine. Fur-
> thermore, the hydrogen-bonding requirement meant that adenine would
> always pair with thymine, while guanine could pair only with cytosine.
> Chargaff's rules then suddenly stood out as a consequence of a double-
> helical structure for DNA. Even more exciting, this type of double helix
> suggested a replication scheme much more satisfactory than my briefly
> considered like-with-like pairing. Always pairing adenine with thymine
> and guanine with cytosine meant that the base sequences of the two in-
> tertwined chains were complementary to each other. Given the base se-
> quences of one chain, that of its partner was automatically determined.
> Conceptually, it was thus very easy *to visualize* how a single chain could
> be the template for the synthesis of a chain with the complementary
> sequence.

THE CONTEXT OF JUSTIFICATION: ARGUMENT PURSUIT

How do scientists decide whether an explanatory hypothesis worth pursuing is worth publishing, a process that may leave them vulnerable to the

devastating criticism of their peers? If scientists are prudent, they will decide to publish only when their hypotheses are well fortified against the possibility of this unpleasant experience. In the context of justification, then, vigilant problem solving turns to the task of transforming the evidence so far gathered into an argument most likely to be persuasive to an audience of peers, an argument that will, in a majority of cases, be built not only out of words, but also out of images.

It is the evidential status that argumentative structures bestow on words and images that permits scientists to claim that a hypothesis is confirmed, that is, that it ought to be accepted, tentatively, as knowledge. There is nothing apodictic about such conclusions, however deductive they may seem. Because all knowledge claims are underdetermined by the evidence that can be adduced in their favor, the door is always open to their modification or replacement: it is always possible that new evidence or a new construal of existing evidence will alter or undermine even a well-established fact or theory, one that virtually everyone in a particular scientific community had hitherto accepted. It is underdetermination, an entailment of induction, that explains the lack of finality in science.

The possibility of rejection when measured against the bar of evidence necessitates that a dividing line be drawn between any virtues a hypothesis may exhibit and the conditions that will ensure its long-term survival. While openness to confirmation or disconfirmation is a virtue, actual confirmation or disconfirmation is not; it is entirely a matter of proof: the cogency of arguments in its favor and the strength of the evidence in its support. Over time, the continued success of a hypothesis in resisting disconfirmation leads, in William Wimsatt's terms (2007), to "generative entrenchment": a hypothesis is transformed into a fact or theory that seems more and more to be beyond question. Such continued resistance is especially recalcitrant when the argumentative and evidential base in its support varies significantly. For example, the arguments and evidence in favor of continental drift form a suite; they derive from geography, climatology, and biogeography (drift's effects) and from seafloor spreading (drift's cause). More recently, an argument that established the age of a Stone Age flute was confirmed—decisively—by a wide variety of dating techniques (Conard, Malina, and Münzel 2009; Higham et al. 2012). As a consequence of this support, the flute became an index of a hitherto unsuspected expansion in the cultural horizon of early humans. It may possibly be objected that variety of confirming evidence is a virtue only, that it lacks probative value. It does, in the context of discovery. But in the context of justification, the more evidence adduced from a variety of sources, the more persuasive an argument is to fellow scientists. In these cases, variety of confirming evidence functions in a manner parallel to prediction, also a virtue in the context of discovery. In the context of justification, fulfilled predictions are persuasive, especially when accompanied by precision.

While confirmation is important, we must not assume that disconfirmation necessarily leads to rejection. Disconfirmation did not follow from Lord Kelvin's physics-based argument that the earth could not be old enough to accommodate biological evolution or from the failure of Newton accurately to predict the perihelion of Mercury. Moreover, disconfirmation is not always applicable. Not all explanations are falsifiable: the theory of punctuated equilibrium is not. Moreover, not all science involves explanation: Carl Scheele's discovery of chlorine and Röntgen's of X-rays are not explanatory, but existential claims. In such cases, especially when disputes arise, it is replication that must come to the rescue. It is only the failure of replication that can differentiate between Röntgen's X-rays and Prosper-René Blondlot's N-rays, a hypothesized form of radiation initially confirmed by other researchers but later discredited (Langmuir 1989). Replication is a form of the practical syllogism: it says that if you do such-and-such, you will produce so-and-so: X-rays or N-rays. Initially, failure to replicate may be attributed to the replicator; repeated failure shifts suspicion to the phenomenon itself. In the case of existential hypotheses, replication is an argument of the first resort because no other is available; in the case of explanatory hypotheses based on experimental findings, it is an argument of the last resort, employed only when the experimenter's competence or honesty is in question. Successful replication, while it does not alter the logical force of Hume's argument against induction, erodes its persuasiveness in particular cases.

Although a lovely hypothesis with wide explanatory scope may survive for a time despite well-confirmed evidence against it, ultimately, not even the loveliest hypothesis can survive repeated or decisive disconfirmation. In the 1950s, the lovely hypothesis for a comma-free genetic code devised by Francis Crick and others held sway—though lacking experimental confirmation—because of its many explanatory virtues: in particular, elegance, visualizability, openness to disconfirmation, unification of existing theories, and fit with existing background beliefs (Crick 1988, 99-101). As Carl Woese wrote,

> The comma-free codes received immediate and almost universal acceptance. . . . They became the focus of the coding field, simply because of their intellectual elegance and the appeal of their numerology. . . . For a period of five years most of the thinking in this area either derived from

the comma-free codes or was judged on the basis of the compatibility with them. (quoted in Hayes 1998, 12)

In the 1960s, Crick's elegant hypothesis was replaced by the current genetic code, which is much less elegant but supported by strong confirmatory evidence that the comma-free codes lacked. The decisive failure of confirmation trumps any virtues a hypothesis may possess—elegance included.

What do confirmative arguments in science look like? We will take as our example Meselson and Stahl's argument concerning the means by which DNA reproduces. This was a problem Watson and Crick left in the wake of their great discovery. In their article, Watson and Crick were appropriately cautious concerning their commitment to the hypothesis they suggested: "The previously published X-ray data on deoxyribose nucleic acid are insufficient for a rigorous test of our structure. So far as we can tell, it is roughly compatible with the experimental data, but it must be regarded as unproved until it has been checked against more exact results" (1953a, 737). Despite this caution, Watson and Crick speculated that "the specific base pairing we have postulated immediately suggests a possible copying mechanism for the genetic material." In a subsequent paper, they are specific about the nature of this copying mechanism, later known as semiconservative replication:

We imagine that prior to duplication the hydrogen bonds are broken, and the two chains unwind and separate. Each chain then acts as a template for the formation onto itself of a new companion chain, so that eventually we shall have *two* pairs of chains, where we had only one before. Moreover, the sequence of base pairs will have been duplicated exactly. (1953b, 966; emphasis in the original)

This was not the only scheme of duplication in contention. Gunther Stent hypothesized a "conservative" mechanism, one in which a new strand of RNA would develop within the physical confines of the DNA double helix. When fully formed, this strand would separate and combine with another of RNA. Together, these two strands would form a template that would stamp out other DNA double helices. Max Delbrück suggested another scheme, dispersion, in which new double helices would form when existing helical chains broke and then rejoined at "growth points." In the chains thus produced, new and existing DNA would alternate. Delbrück also made an important contribution by suggesting a test that would differentiate decisively among the three hypotheses: if the parent strains could be labeled in some way, the daughter strains would exhibit a clear differential distribution.

Four years after this suggestion was made, it was realized in an experimental program, famously referred to as "the most beautiful experiment in biology" (Holmes 2001). After rejecting various alternative labelings, two young scientists, Meselson and Stahl, incorporated a heavy isotope of nitrogen, N^{15}, into the DNA of the bacterium *Escherichia coli*. After fourteen generations, the *E. coli* with its heavier DNA was transferred to a medium in which *E. coli* incorporated standard nitrogen, N^{14}, with its slightly lower atomic weight. The hope was that the distribution of labels and, therefore, of weights, in the contrasting generations of daughter molecules would decisively differentiate among the three hypotheses. Only in the case of Watson and Crick's hypothesis would Meselson and Stahl obtain daughter generations of purely hybrid (N^{14}/N^{15}) DNA molecules, followed by daughter generations consisting of an equal mix of hybrid and N^{14} DNA molecules. In contrast, Stent's hypothesis predicted the presence of N^{15} DNA throughout succeeding generations; Delbrück's predicted neither hybrids, nor pure N^{15}, nor pure N^{14} DNA in succeeding generations.

These differences in atomic weight are minute; they could not be detected even with the most delicate of balances, but only by means of an ultracentrifuge, a machine that spins at an incredible 44,770 revolutions per minute. This is the equivalent of as much as 289,000 times the force of gravity. During centrifugation the DNA is concentrated in the solution and forms a band whose position depends on its weight. As a consequence of this density-gradient centrifugation, after a day or two DNA molecules suspended in a solution of a heavy cousin of sodium chloride, cesium chloride (CsCl), are separated according to their weight. Meselson and Stahl visualized the result for further analysis by use of an automatic camera (fig. 4.3).

In the first ten photographs in figure 4.3*a* sequence (generations 0 to 4.1), we see the stages of bacterial growth from generation to generation. The one dark band in the top photograph represents pure N^{15} DNA. In the second photograph, another, lighter density gradient (hybrid N^{14}/N^{15} DNA) appears to the left of the first (N^{15} DNA). In the third, the initial density gradient begins to fade; in the fourth, the first daughter generation, it completely disappears, and the hybrid alone remains. As we move toward the second daughter generation—photographs 5 through 8—a new, lighter density gradient appears to the left and grows in intensity, indicating the presence of hybrid N^{14}/N^{15} and N^{14} DNA alone. In photographs 9 and 10, this has become the dominant density gradient. The graphs in figure 4.3*b* depict the same sequence in a different visual modality, where the peaks represent the concentration of DNA in the bands.

Figure 4.4 interprets the photographic results of figure 4.3 as a flowchart

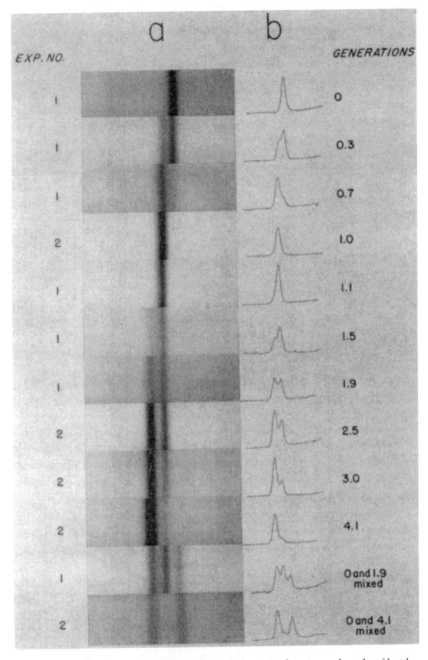

Figure 4.3. Photographs of DNA bands formed after centrifugation and graphs of band density (Meselson and Stahl 1958, 312). On the left side of this figure (a), we see DNA bands showing density gradients for the various "generations" listed in the right-hand column. Density increases as we move from left to right. On the right side (b), we see microdensometer tracings of these DNA bands. The height above the baseline of these tracings corresponds to the concentration of DNA.

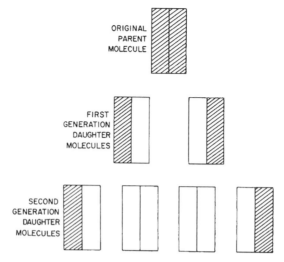

ORIGINAL
PARENT
MOLECULE

FIRST
GENERATION
DAUGHTER
MOLECULES

SECOND
GENERATION
DAUGHTER
MOLECULES

Figure 4.4. Schematic representation of the conclusions drawn in the text from the data presented in figure 4.3, conclusions supportive of the Watson-Crick model (Meselson and Stahl 1958, 318).

(space as a metaphor with time flowing downward) compatible with the Watson-Crick hypothesis: that is, the two components of the parent DNA molecule each form a hybrid DNA molecule after one generation (one component from the parent combined with a new component), then in the second generation each of the two DNA molecules again replicate to form a hybrid DNA molecule plus a new DNA molecule. *Interpret* is the key word. Although figure 4.4 is isomorphic with the photographic evidence, it is not the only isomorphism possible. It is, however, the isomorphism that in the event most scientists would judge the most probable.

Meselson and Stahl's experiment was designed to provide evidence for decisively selecting one among three hypotheses in contention: Watson and Crick's, Stent's, or Delbrück's, each predicting a particular experimental result. The result obtained supported the Watson-Crick hypothesis, and undermined the hypotheses of Delbrück and Stent. The eliminative induction by which they argue can be modeled as a deductive process:

If hypothesis A is the case, x will be the result.
If hypothesis B is the case, y will be the result.
If hypothesis C is the case, z will be the result.
The result is x.
Therefore, A is the case.

The apodeictic quality of this conclusion, however, is an illusion. There is no guarantee that the correct hypothesis is among those being tested, that another hypothesis, also compatible with the experimental evidence, is really the correct choice. Second, there is always the possibility that equipment whose proper functioning has been taken for granted has, in fact, malfunctioned; perhaps the ultracentrifuge was improperly calibrated. Third, it may be the case that one of the procedures by which Meselson and Stahl obtained or manipulated their DNA introduced a variable of which they were not aware, one that compromised their result. In a word, the deductive conclusions Meselson and Stahl derive from their experiment are no stronger than their inductive basis.

Meselson and Stahl were well aware of the tentativeness of their conclusions. Their private thoughts, as revealed by Frederic Holmes (2001), indicate that they would have liked to conclude that their experiment confirmed not only Watson and Crick's hypothesis of semiconservative DNA duplication, but that in doing so, it confirmed the correctness of the Watson-Crick model itself. They would have liked to reason as follows: Watson and Crick inferred a mechanism of duplication from their proposed model, a mechanism our experiment confirms. Therefore, our experiment confirms the model itself. Their figure 4.5 seems to support this inference from the experiment to the truth of the Watson-Crick replication mechanism.

The analogy between figures 4.4 and 4.5 is visually compelling. Nevertheless, Meselson and Stahl (1958, 319) take back in words the inference their analogous images support and encourage:

> The results of the present experiment are in exact accord with the expectations of the Watson-Crick model for DNA duplication. However, it must be emphasized that it has not been shown that the molecular subunits found in the present experiment are single polynucleotide chains or even that the DNA molecules studied here correspond to single DNA molecules possessing the structure proposed by Watson and Crick.

Although a subsidiary experiment (322) conclusively rules out Delbrück's hypothesis, the equivalence of the subunits to Watson-Crick chains, the alleged equivalence of figures 4.4 and 4.5, remains an open question. Moreover, this was not the limit of Meselson and Stahl's skepticism, as their conclusion demonstrates:

> The results presented here give a detailed answer to [the question of the distribution of parental atoms among progeny molecules] and simultaneously

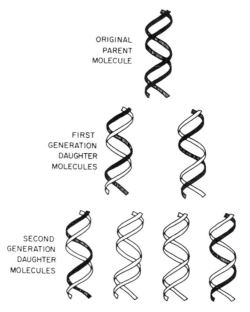

Figure 4.5. Illustration of the mechanism of DNA duplication proposed by Watson and Crick (Meselson and Stahl 1958, 319). Each daughter molecule contains one of the parental chains (black) paired with one new chain (white). Upon continued duplication, the two original parent chains remain intact, so that there will always be found two molecules, each with one parental chain.

direct our attention to other problems whose solution must be the next step in progress toward a complete understanding of the molecular basis of DNA duplication. What are the molecular structures of the subunits of *E. coli* DNA which are passed on intact to each daughter molecule? What is the relationship of these subunits to each other in a DNA molecule? What is the mechanism of the synthesis and dissociation of the subunits in vivo? (322)

Meselson and Stahl's strategic skepticism relates only to *their* science, not to science itself. At the same time that they seem to deny that they have answered the questions they set out to answer, they express an implicit confidence that the progress of science will provide the answers they, apparently, failed to provide. In private, Meselson and Stahl felt as strongly as Watson and Crick that the majority of their fellow scientists would agree that their solution was beyond criticism. From the point of view of these two young scientists, then, the situation could not be more satisfactory: the relevant scientific community would assume full responsibility for a result whose impact on the state of knowledge Meselson and Stahl deliberately minimized.

FROM DISCOVERY TO JUSTIFICATION: ARGUMENT CONSTRUCTION

In the process of justifying a discovery before an audience of peers, scientific argument enables and enacts a series of semiotic transformations. We will illustrate these by means of an analysis of "*Nassarius kraussianus* Shell Beads from Blombos Cave: Evidence for Symbolic Behaviour in the Middle Stone Age," an archaeological article by d'Errico, Henshilwood, Vanhaeren, and van Niekerk (2005). The apparently trivial claim of the title that these shells are beads is, in fact, of vital importance to the science: it concerns the origin of humans who are not only anatomically but also culturally modern. Because beads have a symbolic function, they serve for d'Errico et al. as a surrogate for language, an essential characteristic of advanced hominids. Thus their identification will support the far more general claim that the emergence of culturally modern humans was a gradual process that began in southern Africa about 75,000 years ago.

The expository problem d'Errico and his coauthors must tackle is, how do they justify the claim that a set of tick shells found in a cave is 75,000 years old and was used as an ornament by early humans? Early in their article, they make their argumentative strategy explicit, a degree of explicitness that is a direct result of criticism of a brief report in *Science*. This criticism had questioned the evidential basis of the claim that the forty-one tick shells discovered in the Blombos Cave were, in fact, beads (Henshilwood et al. 2004; Holden 2004):

> Determining whether the *N. kraussianus* shells from the c. 75 ka [about 75,000 years ago] levels at BBC [Blombos Cave] are MSA [Middle Stone Age] beads requires evidence for: i) human agency in their selection, transport and accumulation; ii) manufacturing and/or use wear, and iii) absence of contamination from the LSA [Late Stone Age] layers. Here we demonstrate human involvement in the shells' collection, perforation and use as beads, and provide absolute evidence that the beads derive from the MSA levels. (d'Errico et al. 2005, 10)

When this argument is fully articulated, they aver, it will successfully support its claim as against rival claims. It will undercut the claim that culturally modern humans did not emerge until about 40,000 years ago and in Europe, and the equally mistaken claim that they emerged in Africa, but only about 50,000 years ago and as a consequence of a sudden brain mutation.

Figure 4.6, a set of maps published by d'Errico et al., supports the opening move in their verbal-visual argument. It establishes the location of the forty-one tick shells the archaeologists have discovered. This is not just a location in space per se, but in geological and archaeological space: it is a location that also confirms the age of the shells.

The two spatial maps (a and b) and the spatiotemporal stratigraphic section (c) form a nested set, a sequence of differing iconic representations that zeros in on the M1 layer of the Middle Stone Age stratum of the dig where the forty-one perforated shells, the subject of the article, were found. In map a, the superimposed grid consists of three nested rectangles, the smallest of which, labeled "enlarged," outlines an area of the largest rectangle, the segment of the South African coastline that contains the Blombos Cave dig whose finds are the article's subject. The superimposed grid of the largest rectangle consists of axes calibrated by longitude and latitude and measured by means of an enclosed kilometer scale. It shows us the locations of the Blombos cave and Die Kelders archaeological site near Cape Town.

Map b depicts the area of the Blombos Cave itself; its superimposed grid designates the various dig sites and provides us with a meter scale, with two deictic arrows pointing, respectively, to the north and to the shoreline (linking maps a and b). A legend is enclosed in a small rectangle on the lower right. Map c is a stratigraphic representation; it rotates one of the squares of the Blombos Cave grid in b, H6, ninety degrees, to reveal its archaeological detail, classifying each layer of debris designated by a combination of capital letters. Its grid consists of two parallel vertical lines, the left giving the depth in meters; the right, the archeological era of that depth, from Late to Middle Stone Age. Within each successive map from a to c, then, we zero in more closely on the location of study.

These maps and their accompanying text unite in a common semiotic task. While only the maps can depict the location of the cave, and only its stratigraphy can perform as an index of the age of the tick shells, only words can pinpoint its precise location and place it in its relevant geological context:

> Blombos Cave (BBC) is located 300 km east of Cape Town at 34°24.857′S, 21°13.371′E on the southern Cape shoreline of the Indian Ocean [fig. 4.6a]. The cave is 34.5 m above sea level, some 100 m from the ocean, and formed during Plio-Pleistocene wave cutting of the calcarenite cliff that lies above a basal layer of Table Mountain Sandstone of the Cape

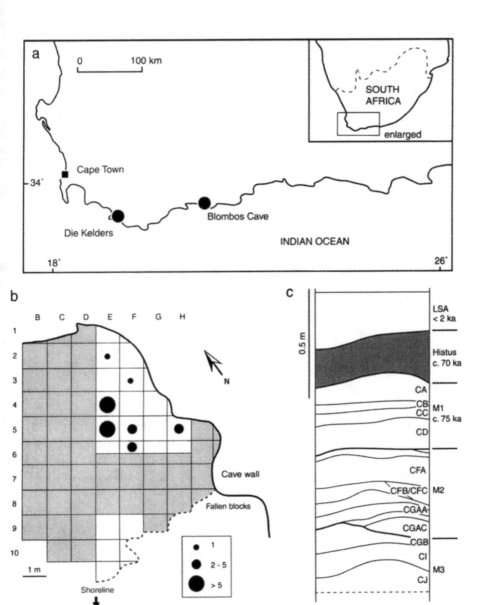

Figure 4.6. Maps of (a) South African coast showing locations of the Blombos cave as well as a nearby site (Die Kelders) where beads were also found; (b) excavated cave area (white) indicating density of MSA (Middle Stone Age) shell beads per square meter; and (c) stratigraphy of square H6 in figure 4.6b (d'Errico et al. 2005, 7).

Supergroup. Deposits within the cave extend over about 50 m² with a further 18 m² of deposit forward of the drip line. (6)

This sequence of words and images establishes that, whatever the origin of these perforations, whether natural or human, the shells are, ceteris paribus, approximately 75,000 years old, it having been determined by optically stimulated luminescence that the layer in which they have been found derives from that era. This is the first step in the verbal-visual argument by d'Errico et al. If it can be established that these shell perforations are human piercings, if it can be proven that their purpose was to turn shells into beads, then, 75,000 years ago on the southern African coast, their manufacturers and wearers exhibited a trait, symbolic communication, plausibly identifiable with behaviorally modern humans. It is thus that these maps and their accompanying stratigraphic section initiate a systematic semiotic transformation from the iconic to the indexical, from the discovery and description of a cache of perforated shells to their role as evidence for the presence of behaviorally modern humans in the Middle Stone Age.

The next step in the argument is the demonstration that the forty-one perforated tick shells exhibit characteristics that make them uniquely suitable for human piercing. This is the joint task of figure 4.7 and the table we reproduce as figure 4.8.

Figure 4.7 allows readers to examine what the authors examined, but arranged in a way to support the authors' argument. It consists of a grid of two parallel rows separated by a horizontal line. In the top row, we see a series of irregular masses among which there are family resemblances; within each mass, we see a varying pattern of light and dark. In addition, we notice

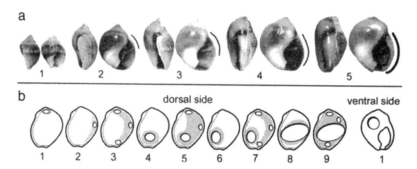

Figure 4.7. (a) Photographs of modern juvenile (1), subadult (2–4), and adult (5) *Nassarius kraussianus* shells from the Duiwenhoks estuary in Cape Province, South Africa; (b) diagrammatic drawings of perforation types on *N. kraussianus* shells from modern and archaeological samples (d'Errico et al. 2005, 11).

Frequency of perforation types (see Fig. 3b) on the dorsal and the ventral side of *Nassarius kraussianus* shells from modern and archaeological assemblages

Nassarius kraussianus assemblages	n	Dorsal side (%)									Ventral side (%)
		1	2	3	4	5	6	7	8	9	1
Modern											
Duivenhoks*	728	4.9	2.9	1.5	1.4	0.7	1.9	2.7	3.8	2.9	51.5
Duivenhoks†	183	1.1	4.9	–	0.5	0.5	1.6	2.2	3.8	3.3	89.1
Goukou*	925	3.4	3	1.1	2.9	1.2	4	3.4	4.6	7	37.7
Goukou†	751	0.3	0.1	–	–	–	–	–	–	–	–
Archaeological											
BBC MSA	41	–	–	–	–	–	63.4	9.8	24.4	2.4	–
BBC LSA	1003	–	0.2	–	0.2	–	0.8	0.1	75.5	7.3	0.6
DK LSA	1095	–	–	1.5	17	5.4	10.4	0.4	35.3	30	3.5

DK = Die Kelders
* modern thanatocoenosis hand gathered throughout the estuary
† modern thanatocoenosis found accumulated at a single spot
‡ hand gathered living population

Figure 4.8. Table arranging data acquired on bead types in figure 4.7*b* (d'Errico 2005, 11).

adjacent black curves of varying size and thickness to the right of four of the masses. In the bottom row, we perceive a series of diagrammatical representations, another series of irregular shapes in each of which are one or more encircled areas of varying size; in addition, shading encompasses areas of varied size. It is words—the labels, legend, and text—that permit us to transform these images into propositions specific to the case at hand. As a result of scanning and matching, we can say of the photographed mollusk shells in dorsal and ventral view that *N. kraussianus* adults are distinguished from subadults and juveniles by their size and by the thickness and expansion of the ventral lip, a thickness the authors imitate and emphasize by means of black curves adjacent to the photographed shells. We can also determine that the inner circles of the diagrammatical representations stand for perforations, and that their shaded areas symbolize the degree of variation in perforation size and shape. Finally, we can say of these representations that only types 6 and 8 exhibit a unique perforation near the lip with a size and location that might possibly be the result of human puncturing.

These sequences demonstrate a principled division of labor between two kinds of iconic representation. Only photographs can conclusively demonstrate the relative size and thickness of lip of the various life stages: juvenile, subadult, and adult. Only diagrams can illustrate the variability in perforation dimensions. From the photographed mollusks, it is a plausible inference that the adult stage 5 in figure 4.7*a*, the stage of all of the shells found at the MSA level, is the best candidate for piercing; from the diagrammed perforations, it is a plausible inference that shells of types 6 and 8 in figure 4.7*b* are the most likely to have been perforated by human activity;

at this point, however, it is equally plausible that their variability was the result of wear.

The words and numbers in the table (our fig. 4.8) support figure 4.7 (d'Errico et al.'s fig. 3), to which the table refers. Like figure 4.7, the table furthers the claim that *Nassarius kraussianus* shell beads from Blombos Cave constitute evidence for symbolic behavior in the Middle Stone Age.

While they are visual displays, tables are not visuals; rather, they are a means of using the resources of the printed page to assist in the clarification of meaning by means of scanning and matching; they are of a piece with such other means of clarification as punctuation and paragraphing. Communicatively, tables are unique only in their systematic use of both dimensions of the printed page—vertical and horizontal. They are understood by recognizing their grids as guides to their symbolic data elements. The grid labels enable the transformation of these elements into propositions. As discussed in chapter 2, tables are a way of arranging their data elements—their words or numbers—so that many parallel propositions (and their corresponding sentences) can be efficiently generated by this series of operations. In the table reproduced in figure 4.8, for example, the bottom three rows, matched against column 6, give us the following:

1. 63.4% of *Nassarius krausianus* shells at site BBC MSA have Type 6 perforations;

2. 0.8% of *Nassarius krausianus* shells at site BBC LSA have Type 6 perforations; and

3. 10.4% of *Nassarius krausianus* shells at site DK [Die Kelders] LSA have Type 6 perforations.

It is this parallelism that permits the ready aggregation of these propositions into such generalizations as "The percentage of *Nassarius krausianus* shells with Type 6 perforations at these archaeological sites varies from 0.8 to 63.4." This parallelism also enables the highlighting of significant differences: for example, the above three propositions generate the genuine news that 63.4% of the recovered shells from the Middle Stone Age Blombos Cave assemblages have unique medium-sized perforations near the lip, as seen in type 6 in figure 4.7*b*. Recourse to the data in figure 4.8 and the pictures in figure 4.7*b* then licenses d'Errico and his coauthors' (2005, 11) generalization:

N. *kraussianus* shells with keyhole perforations on the dorsal side do not exceed 0.4% in living and 30.6% in dead populations [fig. 4.8, two "Goukou" rows]. These perforations vary considerably in number, size, and location [figs. 4.7b and 4.8]. In contrast, all recovered MSA shells are perforated, and 88% have unique medium size perforations located near the lip or larger perforations extending from the lip toward the right edge [types 6 and 8 in fig. 4.7b and fig. 4.8, "BBC MSA" row]. Both of these perforation types are absent in living populations [fig. 4.8, bottom "Goukou" row] and only account for between 5.4% and 8.6% in modern thanatocoenoses [fig. 4.8, types 6 and 8, "Duivenhoks" and "Goukou" rows].

Only a table—words and numbers in combination—could enable the recovery of the exact percentages that license this generalization, and only this generalization could underline so effectively the striking difference that characterizes shells of types 6 and 8.

At this juncture, the argument by d'Errico et al. has progressed to the following point: the shells found in the Middle Stone Age layer—shells that are, ceteris paribus, 75,000 years old—differ characteristically from comparable modern and Late Stone Age populations. Moreover, it is in those very characteristics that their suitability as material for beads lies. We have moved another step in the direction of establishing the article's central claim. It remains to show that the shells exhibit a unique pattern of perforations that is most easily explained by human intervention and to eliminate the possibility that the shells have slumped down from later, Late Stone Age strata.

The perforations of the shells of types 6 and 8 suggest the possibility of human intervention, and the variations in their perforation, patterns of wear. Nevertheless, whether humans actually manufactured beads from these shells given the tools that might have been available, and whether there are on the forty-one shells in question actual signs of wear are questions the positive answers to which would strengthen the central claim of d'Errico et al.

The set of photographs in figure 4.9 depicts an experiment designed to see whether the shells in question could have been pierced by available MSA tools: a stone ("lithic") point, a crab claw, and a bone awl are the possibilities investigated. The goal of the operation is a symmetrical aperture with micro-chipping on the outer prismatic layer—a configuration "similar to that present on most of the MSA shells" (d'Errico et al. 2005, 13).

This figure is divided into panels, each exhibiting the results of an attempt by the authors to pierce the shells, which are photographed in dorsal and ventral views and in various states of magnification, indicated by a scale bar on their lower edges. No matter what the view, the features of the hole are the center of visual attention: in each case, these count as evidence

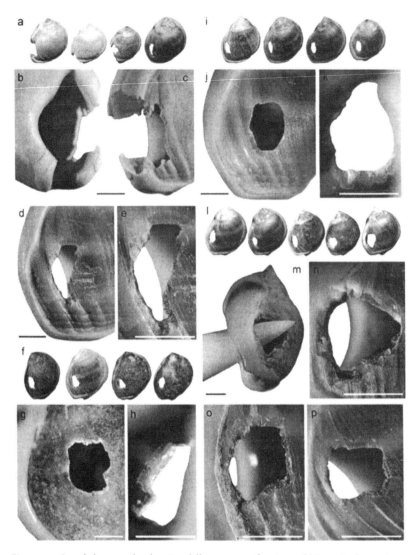

Figure 4.9. Set of photographs showing different magnifications of *Nassarius kraussianus* shells experimentally pierced by various methods: through the aperture with a lithic point (a–e), from the outside with a lithic point (f–h), from the outside with a bone point (i–k), and through the aperture with a bone point (l–o) and a crab claw (p) (d'Errico et al. 2005, 17).

for success or failure in imitating type 6 and 8 perforations, a semiotic trans-formation from the iconic to the indexical.

As interpreted in the text, panels *a* through *e* and *f* through *k* illustrate a series of failures: a stone point that penetrates through the aperture of a shell, and a bone awl or a crab claw that penetrates from the shell's exterior. These methods produce either breakage, unacceptable irregularity, or unac-ceptable micro-chipping of the internal edge, a feature not seen in the hy-pothetically human-punctured shells in question. Panels *l* through *p* record success: we see the bone awl, elliptical in section, penetrating through the aperture, making a hole consistent with the evidence of these shells. A crab claw was equally successful at this task. Read in their designated order, from *a* to *p*, these panels count as evidence for the conclusion that the per-forations in question could be of human origin: the perforations in the shells have become a semiotic index of human intervention.

Having established the possibility that the perforations were made by humans or protohumans, the authors still must establish that this action was taken to create necklaces or bracelets. To do so, text and photographs must work together to show that the pierced shells exhibit patterns of wear consistent with such use. Figure 4.10 and its accompanying text accomplish this purpose. Figure 4.10 consists of a sequence of four panels, each with the same structure: a shell in dorsal view and two close-up views of the shell's hole, the second more highly magnified than the first, as indicated by their scale bars. Each panel displays a different shell: two from the Middle Stone Age, one from the Late Stone Age, and one from the Goukou estuary near the Cave. In each case, the focus is the same: the condition of the parietal wall and the lip close to the anterior canal, marked in the first two pan-els (images *b* and *e*) by deictic arrows. The contrast between the first and last two panels is telling: use-wear patterns are totally absent in the latter two panels.

It is now possible to generalize concerning perforation types 6 through 9 in figure 4.7*b* and figure 4.8. Together, these constitute 100% of the forty-one tick shells. Types 7 to 9 differ from type 6 only in the size of their per-forations. What accounts for the enlarged perforations in these latter cases? Only text can make this conjecture: "a likely explanation for these large perforations and localized wear is that they are the consequence of a gradual enlargement of perforations of Type 6 due to the prolonged use of shells as threaded beads" (2005, 15). D'Errico and his coauthors' justification for their claim is now nearly established. The forty-one tick shells have be-come beads, a transformation from one sort of iconicity to another. As a result of this transformation, two characteristics have become indexical—

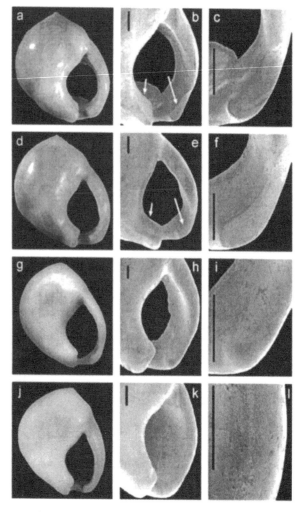

Figure 4.10. Set of electron microscopic photographs of *Nassarius kraussianus* shells
from Middle Stone Age (a–f) level at Blombos Cave, Late Stone Age (g–i) level at Blombos
Cave, and modern shells (j–l) (d'Errico et al. 2005, 18). Scale bars = 1 mm.

the holes themselves, and the signs of wear in the holes—both indices of
human intervention.

We must still determine that these beads have not slumped down from
the Late Stone Age layer. If this were the case, the shells would still be
beads, but they would no longer represent culturally human behavior of an
age coincident with the claim of early origin. Figure 4.11 is integral to an
argument that eliminates this possibility.

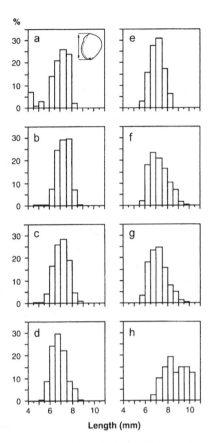

Figure 4.11. Graphs representing distribution of *N. kraussianus* shell lengths in living, dead, and fossil populations: "(a) living population collected by sieving at the Duiwenhoks estuary; (b) hand-gathered living population from the Goukou estuary; (c) hand-gathered modern thanatocoenosis [dead organisms] from Duiwenhoks estuary; (d) hand-gathered modern thanocoenosis from Goukou estuary; (e) accumulation of *N. kraussianus* shells drilled by *Natica tecta* from Duiwenhoks estuary; (f) archaeological *N. kraussianus* from the LSA at Die Kelders Cave; (g) archaeological *N. kraussianus* from LSA at Blombos Cave; (h) archaeological *N. kraussianus* from the MSA at Blombos Cave" (d'Errico et al. 2005, 21).

These histograms represent data trends with regard to the length distribution of *N. kraussianus* shells in living, dead, and fossil populations, as measured by a common standard depicted in the upper right-hand corner of the first histogram. The Gestalt principles of similarity and contrast allow us to perceive the varying length of each of the bars; at the same time, proximity allows us to perceive all eight of the histograms as a set of sets. In this more encompassing context, similarity and contrast lead to the perception that histogram (h) is anomalous. With the aid of labels, legend, and text, readers move from patterns to propositions. For example, in histogram (a) the first bar represents the fact that approximately 7% of shells are 4 to 4.5 mm in length. More important, induction across the eight histograms allows readers to infer that collections of Modern and Late Stone Age shells are essentially similar (bar graphs *a* through *g*); however, at bar graph *h*—a Middle Stone Age collection—the hypothesis fails: while all other populations of *N. kraussianus* vary from 4.5 to 9 mm in length, MSA shells of the same species vary from 6.5 to 10.5 mm. This difference allows d'Errico et al. to rebut any criticism that these shells may have slumped down from Late Stone Age layers. Their current location is indeed an index of their age.

The most plausible conclusion from this interaction between the visual evidence and the text is now clear: approximately 75,000 years ago the forty-one shells were turned into beads, strung into bracelets and necklaces, and worn by human or protohuman beings inhabiting what is now the southern coast of South Africa. This conclusion also allows us to infer that we have in our possession "an unambiguous marker of symbolically mediated behavior" among these inhabitants (19). We can make this inference because "beadwork represents a reliable proxy for acquisition of language and fully modern cognitive abilities" (20). What is language, after all, but a sophisticated system of symbolic interchange? In the course of this long argument, the tick shells in question have been transformed within Peirce's iconic category: they have become decorative beads. As a consequence of this verbal-visual argument, the iconic has also been transformed into the indexical: the beads have become evidence for modern human behavior. Finally, the indexical has been transformed into the symbolic: the beads point in the direction of language and, therefore, of modernity in the Middle Stone Age.

Meselson and Stahl's article could also have illustrated the point we have just made. Their reasoning rests squarely on a sequence of symbolic equivalences that begin with the dinner-table "experiment" in which the fingernail stands for the DNA molecule: a metaphoric transformation. In the experiment proper, metonymy and synecdoche rule. The nitrogen stands for all nitrogen; the DNA of *E. coli* stands for all DNA, two metonymic transformations. The weight of the DNA molecule stands for the mechanism of reproduction, a synecdochical transformation, the part standing for the whole. Finally, the interpretation of the photograph we reproduce as figure 4.3 turns an iconic record of an event into an index of its cause as depicted in figures 4.4 and 4.5. In Meselson and Stahl and d'Errico et al., the pattern of transformations varies; the fact that there is a pattern does not.

We suggest that such semiotic transformations are general, the routine result of the operation of scientific argument.

CONCLUSION

Scientific argument is most perspicuously viewed as a dynamic process that begins with the formation of a problem to be pursued and ends with the public statement of a claim supported by reasons and evidence. At any time, this statement can be subject to interrogation, opening a space for new research that may amplify, modify, and overturn previous claims. This process is rational throughout, at least in the sense that at each stage a decision is reached for reasons that, the authors anticipate, a community of scientists would deem acceptable. Moreover, at no stage in this process is the visual excluded from making a contribution of epistemic significance; indeed, in many cases, the visual is epistemically central. While we do not claim that all scientific arguments can be reconstructed like the examples we have chosen, we do claim eliminative induction has a general application to much of science and to many scientific arguments. This is not a claim that the inductions of science achieve absolute certainty. Nothing we say about the rationality of science at all stages affects in the slightest two sturdy philosophical arguments against the ultimate cogency of induction: Hume's and its entailment, the argument from underdetermination. We also contend that, regardless of their structure, by means of verbal-visual interaction, the preponderance of scientific arguments, especially those in the experimental realm, enable and enact a series of significant semiotic transformations with epistemic purchase.

We need to add two important qualifications to our claims. The centrality of rationality in our model does not mean that we ignore the distinction between any virtues that scientific hypotheses may exhibit and the status of these hypotheses as explanations acceptable to science. While we want to say that it is rational for scientists to find these hypotheses attractive just on the basis of their virtues, we also want to say that it is equally rational for them to abandon them when, however virtuous, they fail continually against the bar of countervailing evidence. In addition, our model of scientific argument is just that. It is not intended to reflect the social process of investigation, a process more nuanced in its day-to-day execution than we have portrayed (e.g., Latour and Woolgar 1979; Knorr-Cetina 1981, 1999; Collins 2004; Pauwels 2006).

In our analysis of scientific argument, we have left undiscussed a family of sciences in which time is a central issue. This is not the t of physics equations or of line graphs; this is the time that is history's medium. In these sciences, argument must be transformed into narrative; accordingly, it seems appropriate to call such sciences as geology, evolutionary biology, anthropology, and archaeology "historical." In our next chapter, we contrast the efforts of Charles Darwin and Alfred Wegener to create a convincing geologic narrative with words and images. Darwin was successful; Wegener was not but his successors were. We also attempt to find out why as well as how, with the discovery of seafloor spreading, Wegener's failure was converted to success.

CHAPTER FIVE

Visual Argument and Narrative in the "Historical" Sciences: The Example of Geology

I could a tale unfold whose lightest word
Would harrow up thy soul, freeze thy young blood,
Make thy two eyes, like stars, start from their spheres,
Thy knotted and combined locks to part
And each particular hair to stand on end,
Like quills upon the fretful porpentine.
—William Shakespeare, *Hamlet*

While all sciences have as their goal the discovery of processes that repeat themselves over time, only in such "historical" sciences as geology, evolutionary biology, anthropology, and archaeology is the goal also the explanation of events unique in time, those generated by means of such processes. Geology tells us the story of the earth; evolutionary biology, the story of living things; anthropology and archaeology, the story of humankind. While the historical sciences differ from each other—unlike geology and evolutionary biology, archaeology and anthropology are concerned with motives—all are alike in that they are grounded in evidence procured by methods approved by their disciplinary communities. As a consequence, their narratives differ from those of fiction: their credibility depends on the continued cogency of underlying arguments in which words and images interact to make a coherent argument and tell a coherent story.

Our concern in this chapter is not with ordinary narratives, such as those that present the story behind some scientific discovery, but with "argumentative narratives," structures in which narrative and argument form, as it were, two sides of a Möbius strip. Change over time is critical to both kinds of narrative, but the main focus of the argumentative narrative in the historical sciences is not the time as experienced by the characters and their readers, but "the uniform, qualitatively undifferentiated moments in which all change occurs" (Dauenhauer 2005). The concern is also with lengths of time that far exceed the span of anyone's existence, indeed, of the historical record.

In this chapter, we will contrast the work of a scientist who succeeded in telling a credible story and one who did not. Charles Darwin (1842) successfully attributed creation of fringing reefs, barrier reefs, and atolls to the same cause: subsidence, their sinking into the seas in which they sit. Alfred Wegener (1929; 1966) unsuccessfully attributed the present position of the continents to their drift over time through an underlying permeable base. It can be argued that Wegener's strident advocacy of this form of continental drift in the final edition of *The Origin of Continents and Oceans* was no more than outrageous speculation supported by hollow rhetoric; certainly, this was the opinion of many of his geological contemporaries. We argue instead that this distinguished scientist contributed to his science by keeping alive his intuition that the continents did, in fact, change position over geological time. It is this intuition that the advocates of seafloor spreading successfully supported several decades after Wegener's death.

These two geological narratives—Darwin's concerning reef-atoll systems, and Wegener's and his successors' concerning continents and oceans—are part of the same story: the earth's surface features are shaped by the operation of subterranean and subaqueous forces of enormous power, forces that in general operate slowly but cumulatively over geological time, splitting apart and moving large land masses. These narratives were also constructed in the same way, generated by arguments that issued from a reciprocating combination of induction and deduction. In each case, images were central to the establishment of the conviction that the right answer had at last been found.

DARWIN: ON THE ORIGIN OF CORAL REEFS

In a letter to his older sister, Susan, written toward the end of the *Beagle* voyage, Charles Darwin conveys his excitement about his new-found geological career. But he also expresses a trepidation that will haunt him throughout his life: the persistent anxiety that his theories will not be believed for want of evidence. To guard against this dreaded eventuality, he assiduously collects scientific observations and physical evidence: "This last trip has added half a mule's load," he avers, "for without plenty of proof I do not expect a word of what I have above written to be believed" (Darwin 1835, letter 275).

We know from his *Autobiography* (1887) what we might readily infer from this sentence: these observations and this physical evidence had been gathered with their explanation already firmly in mind. Reflecting on the first publication in the trilogy of geographical monographs (Darwin 1836, 1844, 1846) that were the early fruit of his five-year voyage, he makes clear a bold leap from observations to a hypothesis from whose vantage those observations could count as evidence:

> No other work of mine was begun in so deductive a spirit as this, for the whole theory was thought out on the west coast of South America, before I had seen a true coral reef. I had therefore only to verify and extend my views by a careful examination of living reefs. But it should be observed that I had during the previous two years been incessantly attending to the effects on the shores of South America of the intermittent elevation of the land, together with denudation and the deposition of sediment. This necessarily led me to reflect much on the effects of subsidence, and it was easy to replace in imagination the continued deposition of sediment by the upward growth of corals. To do this was to form my theory of the formation of barrier reefs and atolls. (1887, 98–99)

In a series of papers published during the *Beagle* voyage and soon after his return to England, Darwin lays before his geology-reading public a theory of subsidence and elevation broad enough to explain five tectonic phenomena that might not otherwise be linked: coral reefs, earthquakes, volcanic islands, mountain building, and the elevation of South America over geological time. Published in 1836, the first paper in this series supports this elevation by pointing to the traces of "an ancient channel, which must have traversed a great portion of the southern part of the continent before the elevation of the tertiary groups [most recent, geologically speaking]" (211). Moreover, from the fact that "*recent* shells are *littoral shells* of the neighboring shores" (211; his emphasis), he infers that this elevation must have been gradual. In a paper read in 1837, he attributes this continental elevation, the creation of mountain chains, the incidence of earthquakes, and the eruption of volcanoes to the same cause, the fact that "the crust of the globe in Chili rests on a lake of molten stone" (1838, 656). To illustrate the effects of this continent-wide tectonic underpinning, Darwin singles out the earthquake that destroyed the city of Concepción on the morning of February 20, 1835. A report on the paper he delivered makes this underpinning clear:

In order to enable the reader, who may be more familiar with European than South American geography, to comprehend the vast surface which was affected by the earthquake . . . he stated, that it had a north and south range, equal in extent to the distance between the North Sea and the Mediterranean: that we must imagine the eastern coast of England to be permanently raised; and a train of volcanos to become active in the southern extremity of Norway; also that of a submarine volcano burst forth near the northern extremity of Ireland; and that the long dormant volcanos of the Cantal and Auvergne, each sent up a column of smoke. (656)

In the same year, 1837, Darwin read a paper before the Geological Society showing that the three known classes of reef-island systems—fringing, barrier, and atolls—are formed by the same force, subsidence. He says that "the land with the attached reefs subsides very gradually from the action of subterranean causes, the coral building polypi soon again raise their solid masses to the level of the water; but not so with the land: each inch lost is irreclaimably gone:—as the whole gradually sinks, the water gains foot by foot on the shore, till the last and highest peak is finally submerged" (553). This phenomenon is but one component of the worldwide subsidence and elevation of the earth's crust. In the Pacific and Indian Oceans, "bands of elevation and subsidence alternate" (554). Darwin's talk is illustrated "by the aid of sections" (553), visuals unfortunately not included in the Society's reported summary.

In an extended paper on South American geology, read in 1838, Darwin repeats and expands on the conclusions of the earlier papers: "the facts appear to me clearly to indicate some slow, but in effects, great change in the form of the surface of the fluid on which the land rests" (608). He also ventures for the first time a possible explanation of this process, one borrowed from the geophysicist William Hopkins: "when the crust yields to the tension, caused by its gradual elevation, there is a jar at the moment of rupture, and a greater movement may be produced by the tilting up of the edges of the strata and by the passage of the fluid rock between them" (621). Darwin reproduces a woodcut from Hopkins to illustrate this phenomenon (fig. 5.1). It is not a picture of actual strata; it depicts, rather, a theory of their formation by means of a calculable nexus of forces. It is indexical, referring to causes, rather than iconic, depicting the way the world looks.

Evidence in favor of this theory is provided in a table whose arrangement, perceived by means of Gestalt principles of comparison and contrast

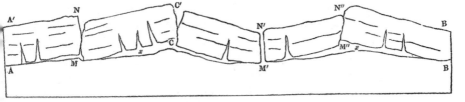

Figure 5.1. Diagram illustrating strata rising due to underlying fluid mass (Darwin 1840, 625).

and analyzed by scanning and matching the rows and columns, highlights the coincidence of events in a wide disparity of latitudes (Darwin 1840):

1835. January 20th	Osorno,	lat. 40°31′S. in eruption.
Before daylight	Aconcagua,	lat. 32°30′S. in eruption.
in the morning.	Coseguina,	lat. 13°N. in terrific eruption, continuing in activity during the ensuing months.

The distance between Osorno and Aconcagua is 480 miles, and between Aconcagua and Coseguina, 2,700.[1] Despite this great distance, Darwin feels it is highly probable that these three events have a single cause. He says that he is "strongly inclined to believe that the subterranean forces manifest their action beneath a large portion of the South American continent, in the same intermittent manner as, in accordance with all observation, they do beneath isolated volcanos—that is, remaining for a period dormant, and then bursting forth throughout considerable districts with renewed vigour" (615).

The worldwide scope of these elevatory forces is exemplified in Darwin's 1839 paper on the parallel "roads" of Glen Roy, a remarkable series of Scottish mountain terraces whose discovery provoked considerable geological controversy in the first half of the nineteenth century. Were they the product of a freshwater lake that once filled the valley or were they marine beaches that had risen considerably above sea level over a vast expanse of time (Browne 1995, 376–78)? In his paper, Darwin makes explicit the analogy to his South American case. In both locations, he avers, the sea has left these terraces as its calling card.[2] After this series of marine incursions, the

---

1. These are Darwin's figures. The actual total distance is around 3,700 miles. The actual latitude of Osorno is 40°06,′ Aconcagua 32°39.′

2. It is irrelevant to our purposes that Darwin's theory of the formation of the parallel roads was in error.

land in Glen Roy was raised to its current level by underlying forces: "the movements appear to have been of the same order with those now in progress in South America; and in that country the elevation of certain wide areas, . . . cannot be attributed to any other cause than an actual *movement* in the subterranean expanse of molten rock" (79; Darwin's emphasis).

In his 1842 monograph, *The Structure and Distribution of Coral Reefs*, Darwin works out the implications of his theory for the creation of one of the most conspicuous geological formations of the Pacific Basin. Although his theory of the formation of these reefs was the result of an abductive leap from a scattering of facts to a bold hypothesis, the book's argument is strictly inductive. Darwin begins with a description of Keeling Atoll so detailed that readers are cast in the role of virtual witnesses who can attest to facts soon destined to be transformed into theory. Below is an example of this technique, a testament to Darwin's meticulousness, a projection in every sentence of the careful researcher who, literally, leaves no stone unturned:

> On the outside of the reef much sediment must be formed by the action of the surf on the rolled fragments of coral; but, in the calm waters of the lagoon, this can take place only in a small degree. There are, however, other and unexpected agents at work here: large shoals of two species of Scarus, one inhabiting the surf outside the reef and the other the lagoon, subsist entirely, as I was assured by Mr. Liesk, the intelligent resident before referred to, by browsing on the living polypifers. I opened several of these fish, which are very numerous and of considerable size, and I found their intestines distended by small pieces of coral, and finely ground calcareous matter. This must daily pass from them as the finest sediment; much also must be produced by the infinitely numerous vermiform and molluscous animals, which make cavities in almost every block of coral. Dr. J. Allan, of Forres, who has enjoyed the best means of observation, informs me in a letter that the Holothuriæ (a family of Radiata) subsist on living coral; and the singular structure of bone within the anterior extremity of their bodies, certainly appears well adapted for this purpose. (14)

Darwin now shifts from verbal to visual witnessing, beginning with a navigator's map. In figure 5.2, Keeling Atoll is foregrounded against a deliberately characterless sea. Dramatically to shift this diagram's orientation, to transform it from geography to geology, Darwin creates figure 5.3 by rotating 5.2 by ninety degrees on its axis, a movement that discloses it in vertical

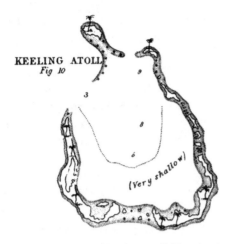

Figure 5.2. A navigator's map of Keeling Atoll (Darwin 1842, plate 1).

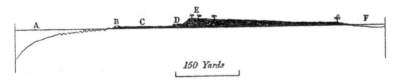

Figure 5.3. A map displaying the structural features of Keeling Atoll (Darwin 1842, 5).

section spanning the highest point of land to 25 fathoms below sea level. At the same time, he simplifies its contours to reveal its essential structural features. In addition, by means of considerable, but constructive, distortion, he clarifies aspects of those features that would be masked by a rendering in true proportions.[3] This simplification and distortion serve as a prelude to theoretical appropriation.

While the diagram permits us to see the island's structural features as a single Gestalt, a task no words could accomplish, it is only words that can identify these and that can, moreover, highlight the crucial role of one particular feature: "The wood-cut represents a section across one of the islets on the reef, but if all that part which is above the level of *C* were removed, the section would be that of the simple reef, as it occurs where no islet has been formed. It is this reef which essentially forms the atoll" (10). The

---

3. "The section is true to the scale in a horizontal line, but it could not be made so in a vertical one, as the average greatest height of the land is only between six and twelve feet above high-water mark" (Darwin 1842, 6). See Browne 1995, p. 271, for Darwin's original sketch.

words and the diagram are closely interwoven, a linkage marked by the coincidence of the letter C, discovered by scanning and matching from text to
image. In creating this spatial representation, Darwin has displaced the atoll
from its geographical context; he has recontextualized it geometrically by
means of a superimposed grid that signals a significant semiotic shift: from
an object in the world of nature to an object in the world of theory. The diagram's geometrical character is now at the center of our field of attention:
the atoll has become its structure, a structure determined by a calculable
nexus of forces.

But can the structure of Keeling Atoll be generalized so as to apply to
all reef-island systems? In pursuit of this goal, Darwin's chapters 2 through
4 detail the similarities among the three classes of reef-island systems—
atolls, barrier reefs, and fringing reefs. This is accomplished by means of
an extensive catalogue and a comprehensive anthology of diagrams whose
comparison is made simpler by the adoption of a uniform visual code.[4] This
catalogue and these diagrams demonstrate that reef-island systems share all
of their essential features, a conclusion reached by scanning and matching
and Gestalt comparison:

> The general resemblance between the reefs of the barrier and atoll
> classes may be seen in the small, but accurately reduced charts . . . [T]his
> resemblance can be further shown to extend to every part of the struc-
> ture. (41) . . . If we look at a set of charts of barrier-reefs, and leave out
> in imagination the encircled land, we shall find that, besides the many
> points already noticed of resemblance, or rather of identity in structure
> with atolls, there is a close general agreement in form, average dimen-
> sions, and grouping. (45)

By the end of the first part of *Structure*, Keeling Atoll has come to stand
for all reef-island systems; it does so by means of a transformation from
the iconic, the representation of one reef-island system, to the symbolic, a
semiotic category by means of which in its essential features one system
stands for all.

Darwin's initial structural model is static, a status emphasized by the
assignment of agency to persons rather than to geological forces. Darwin

---

4. "In the several original surveys, from which the small plans on this plate have been
reduced, the coral-reefs are engraved in very different styles. For the sake of uniformity, I have
adopted the style used in the charts of the Chagos Archipelago, published by the East Indian
Company, from the survey by Captain Moresby and Lieutenant Powell" (Darwin 1842, xvii).

and *Beagle* captain Fitzroy observe and measure; Darwin's correspondents and colleagues Liesk and Allan observe and inform; the earth holds still for its portrait. Throughout the first part of *Structure*, however, by suggesting that the structural features he has just revealed are "the effect of uniform laws . . . that some renovating agency (namely subsidence) comes into play at intervals, and perpetuates their original structure" (24, 31), Darwin anticipates the dynamic model he will soon reveal. These hints foreshadow a transformation that will allow us to reread a passage like the one below as evidence in an argument for a theory of subsidence:

> On the western side, also, of the atoll, where I have described a bed of sand and fragments with trees growing out of it, in front of an old beach, it struck both Lieutenant Sulivan and myself, from the manner in which the trees were being washed down, that the surf had lately recommenced an attack on this line of coast. Appearances indicating a slight encroachment of the water on the land, are plainer within the lagoon: I noticed in several places, both on its windward and leeward shores, old cocoa-nut trees falling with their roots undermined, and the rotten stumps of others on the beach, where the inhabitants assured us the cocoa-nut could not now grow. Captain Fitzroy pointed out to me, near the settlement, the foundation posts of a shed, now washed by every tide, but which the inhabitants stated, had seven years before stood above high watermark. (17–18)

The same facts that in the first part of *Structure* coalesced to build a static model of the reef-island system form in the second part the inductive base of a causal argument for a dynamic theory based on the subsidence of large portions of the earth's crust. In this radical recontextualization, we move from description secured by facts to theory secured by evidence. Despite his theory's actual origin—in a bold analogical leap, a heroic reenvisioning "in imagination"—Darwin understood that it was only by means of a meticulous accumulation of evidence that he could convince his professional peers of its truth, a task to which he devoted five years: "it is very pleasant easy work putting together the frame of a geological theory, but it is just as tough a job collecting & comparing the hard unbending facts" (Darwin 1987, 207).

If his argument was to be given a fair hearing, however, Darwin had to give a fair hearing to competing theories. It was his personal and professional misfortune that his chief competitor was his mentor and friend, Charles Lyell. "The circular or oval forms of the numerous coral isles of the

Pacific with the lagoons at their centre," Lyell had asserted in his magisterial *Principles of Geology*, "naturally suggest the idea that they are nothing more than the crests of submarine volcanoes, having the rims and bottoms of their craters overgrown by corals" (1832, 341). In Darwin's view, Lyell's theory had to be abandoned. It could not explain the presence of fringing nor of barrier reefs; neither could it explain the fact that all reef-island systems were low lying, or that the coral of which they were mainly composed could live only in relatively shallow waters. Finally, the theory was undermined by the general distribution of reef-island systems far from volcanic areas.

Darwin solved his problem by an exercise in diplomacy. While he dismissed Lyell's theory, he did not criticize Lyell himself, who with one exception is mentioned only in favorable contexts, generally as an authority and, in one particular case, an authority on subsidence, the mechanism behind Darwin's own theory: "It is very remarkable that Mr. Lyell, even in the first edition of his *Principles of Geology*, inferred that the amount of subsidence in the Pacific must have exceeded that of elevation, from the area of land being very small relatively to the agents there tending to form it, namely, the growth of coral and volcanic action" (1842, 95; see also 29, 71–72, 118, 137, 143, 175). This strategy succeeded. When Darwin published his theory, Lyell's concurrence was virtually immediate and especially gratifying: "I must give up my volcanic theory for ever," Lyell wrote, "though it cost me a pang at first, for it accounted for so much, the annular [circular] form, the central lagoon, the sudden rising of an isolated mountain in a deep sea" (Darwin 1892, 293). Vital to that acceptance is the fact that Darwin's rival theory is at bottom an application of Lyell's central insight that the earth's current configuration is the result of gradual change over eons of geological time.

Having dealt with and dismissed rival theories, most especially that of his mentor, Darwin devotes the penultimate chapter of *Structure* to his argument that subsidence, the gradual descent of large portions of the earth's crust, accompanied by slow coral growth, causes the transformation from fringing to barrier reefs and from barrier reefs to atolls. The Polynesian island Bora Bora is the exemplar for Darwin's theory of subsidence. An island surrounded by a barrier reef, it is in an intermediate stage between fringing reef and atoll. Readers first encounter Bora Bora in the form of a woodcut, seeing it in figure 5.4 as an explorer might after rising in a hot-air balloon to an elevation that captures both the island's mountain and the reef beyond. Mt. Otemanu, dotted with coconut palms, dominates the scene. Behind the mountain is a placid lagoon. In the background is a barrier reef, surmounted by palm trees. But even in this realistic depiction, Bora Bora's

Figure 5.4. Picture of Bora Bora centered upon Mt. Otemanu, the island's highest point (Darwin 1842, 3). Derived from the voyage of the *Coquille* led by Capt. Duperrey.

recontextualization as a theoretical object has stealthily begun. Darwin says that he has "taken the liberty of simplifying the foreground, and leaving out a mountainous island in the far distance" (1842, 2n).

The next step toward theory is the transformation of Bora Bora from a "realistic" depiction into a map, figure 5.5. To achieve this, Darwin rotates the viewer's perspective by ninety degrees. We are no longer looking at Bora Bora from a long horizontal distance but from a great height, directly above. The illusion of depth in figure 5.4—the product of an artistic code composed of shading and perspective—has been replaced in figure 5.5 by the imposition of a scale.[5] In the interest of placing the reef-island's structure unequivocally in the foreground, the actual has been simplified: some lagoon islets have been omitted, and no attempt has been made to depict the distribution of the island's flora. In line with this purpose, the actual has also been enhanced: the depth of the lagoon in fathoms is variously indicated, and the lagoon is, as it were, drained in order to reveal the contours of its underlying reef. In the realistic rendering, light and shade represent the mountain as a three-dimensional object; in the map, the height of the mountain is symbolized by parallel lines, signaling a change from an artistic to a cartographic code. In the artistic rendering, the size and location of the coconut palms are reproduced so as to reflect their actual size and location; in the map, the repeated coconut palms are only symbols designed to help viewers differentiate the land from the reef below it (215). What is representational in the realistic rendering is symbolized in the map. This uniformity of representation facilitates structural comparisons among reef-island systems, which, in turn, facilitate the transformation of facts on the ground into evidence for Darwin's argument.

5. The scale is presented separately on p. 216 of Darwin's *Structure*.

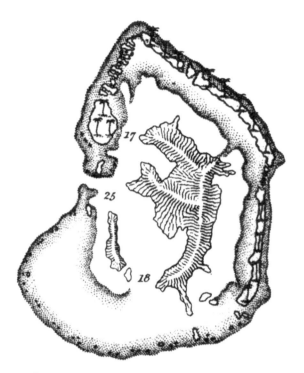

Figure 5.5. Aerial map of Bora Bora from the survey of Capt. Duperrey in the *Coquille* (Darwin 1842, 215). Scale 1/4 inch to a mile. The tinted area shows the extent of the reef. The area that encloses the coconut trees (exaggerated in scale) represents the coral islets. The numbers 17, 28, and 25 are the depth of the lagoon in fathoms of six English feet.

Darwin moves decisively into the realm of theory when in figures 5.6 and 5.7 he transforms Bora Bora into two diagrams that represent changes in spatial relationships over time. To disclose the atoll's vertical section, he rotates the perspective in figure 5.5 by ninety degrees yet again. At the same time, to reveal the essential structural features of this section, structures to be located by scanning and matching, he simplifies the atoll's contours and, by means of considerable vertical distortion, clarifies aspects of those features that would be masked by a rendering in true proportions. Finally, he superimposes a grid so that the reader can view those features through a geometrical lens (99n). In this, he follows the procedure used with the Keeling Atoll (figs. 5.2 and 5.3). In the case of Bora Bora, however, the static has become the dynamic. Figure 5.6 depicts evolutionary succession as a consequence of subsidence; Darwin asks the reader to reproduce this consequence by animating the diagram "in imagination":

Let us in imagination place within one of the subsiding areas, an island surrounded by a "fringing reef"—that kind, which alone offers no difficulty in the explanation of its origin. Let the unbroken lines and the oblique shading in the woodcut [our fig. 5.6] represent a vertical section through such an island; and the horizontal shading will represent the section of the reef. Now, as the island sinks down, either a few feet at a time or quite insensibly, we may safely infer from what we know of the conditions favourable to the growth of coral, that the living masses bathed by the surf on the margin of the reef, will soon regain the surface. The water, however, will encroach, little by little, on the shore, the island becoming lower and smaller, and the space between the edge of the reef and the beach proportionally broader. (98–99)

By means of verbal-visual interaction, figure 5.6 comes to stand for any reef-island system, not Bora Bora alone: it is a dynamic model of reef-island evolution over geological time. In figure 5.6, the anchored boat (to the right of letter C) stands for the depth of any lagoon C; the palm trees (near A'), the existence of land on any coral reef; the differential hatchings, any reef and

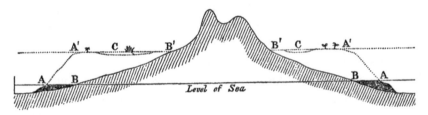

Figure 5.6. Vertical section showing formation of a fringing reef (AA) and a barrier reef (A'A'), based on Bora Bora (Darwin 1842, 98).

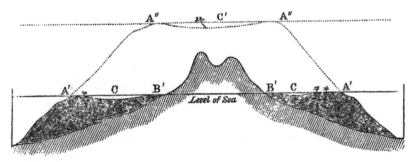

Figure 5.7. Vertical view of Bora Bora showing formation of atoll (Darwin 1842, 100).

its island; the solid line through *AA*, the present sea level; the dotted line through *A'A'*, the future sea level.

In figure 5.7, a sister diagram, we leap ahead in time. By this point, the reef-island system has subsided further still: the barrier reef has become an atoll. In this diagram, the solid line through *A'A'* represents the present; dotted line through *A"A"*, the future, a time when the island will have disappeared below the sea.

In figures 5.6 and 5.7, Darwin has shifted our attention from the appearance of Bora Bora at any one time to the evolution of any reef-island system over time. It is a shift from the static to the dynamic, a transformation from the iconic to the indexical, indicative of a calculable nexus of forces:

> We are now able to perceive that the close similarity in form, dimensions, structure and relative position . . . between fringing and encircling barrier-reefs, and between these latter and atolls, is the necessary result of the transformation, during subsidence, of the one class into the other. On this view the three classes of reefs ought to graduate into each other. (102)

As a consequence of this transformation, Darwin's initial classification of reef-island systems is revealed as a convenient fiction, a concession to the short life span of human beings who, though they cannot see, can by means of argument bring to the forefront of their consciousness the evolution of these systems and the *vera causa*—the true cause—of that evolution over vast tracts of time. In *On the Origin of Species* (1859), in a parallel argument, species will have a status analogous to reef-island systems, species whose *vera causa* is natural selection.

Darwin has not yet completed his theoretical task: geological theories cannot stop at process explanations because geology is a historical science. In the *Voyage of the Beagle*, three years earlier, Darwin views the history of the earth as the interplay of tectonic forces over geological time: "We may thus, like unto a geologist who had lived his ten thousand years and kept a record of the passing changes, gain some insight into the great system by which the surface of this globe has been broken up, and land and water interchanged" (1845, 480). To realize this vision in fact, however, the master argument of *Structure* must be transformed into a master narrative. Unlike their fictional counterparts, such narratives are credible only so long as their underlying arguments hold true. In a historical science like geology, arguments and the narratives inferred from them are epistemologically equivalent.

When Darwin concludes his argument in favor of his theory of subsidence, Keeling Atoll and Bora Bora have been transformed into typical reef-island systems; at the same time, they have been turned from material into theoretical objects, defined by their geometry in relation to their surrounding seas, and characterized by subsidence, the twin of elevation, a calculable nexus of forces. Now what has been decontextualized in the interest of theory must be re-contextualized as narrative under the lens of theory: "the history of [a particular] atoll" can be reconstructed only if the general argument for subsidence is "modified by occasional accidents which might have been anticipated as probable" (1842, 114); that is, only insofar as the geological features of a particular reef-island system are taken into consideration can we imagine what its past might have been, and what its future is likely to be. The particular geological features of New Caledonia, for example, allow us to turn a general process into a specific narrative, to envision the story of a unique future that stems from a unique past:

> if, in imagination, we complete the subsidence of that great island, we might anticipate from the present broken condition of the northern portion of the reef, and from the almost entire absence of reefs on the eastern coast, that the barrier-reef after repeated subsidences, would become during its upward growth separated into distinct portions; and these portions would tend to assume an atoll-like structure, from the coral growing with vigour round their entire circumference, when freely exposed to an open sea. (110)

Whatever applies to New Caledonia applies generally to every reef-island system, a narrative that sweeps across the Pacific from every point on the compass:

> We there see vast areas rising, with volcanic matter every now and then bursting forth through the vents and fissures with which they are traversed. We see other wide spaces slowly sinking without any volcanic outbursts; and we may feel sure, that this sinking must have been immense in amount as well as in area, thus to have buried over the broad face of the ocean every one of those mountains, above which atolls now stand like monuments, marking the place of their former existence. (148)

In addition, Darwin insists on the broader implications of his theory on climate change and, proleptically, "on the distribution of organic beings."

In the final chapter of *Structure*, we are introduced to figure 5.8, a map

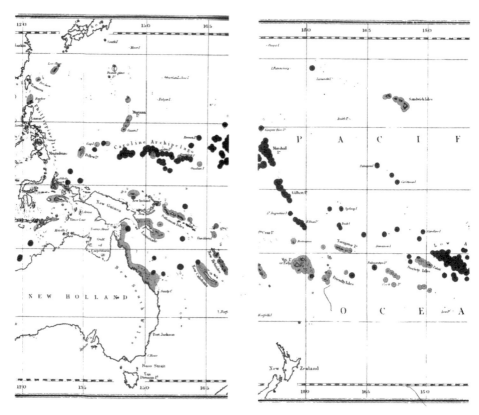

Figure 5.8. Map of atolls, barrier, and fringing reefs of the Pacific (Darwin 1842, plate 3). The scale is such that each square contains 810,000 square miles. Original covers two pages and uses color.

of the Pacific Ocean that indicates, by its differential coloring, the location of every reef-island system and of the "Ring of Fire," the chain of volcanoes at the margins of the Pacific. On the map, those systems colored bright blue designate atolls and lagoon islands; pale blue, barrier reefs; red, fringing reefs; vermillion, active volcanoes.[6] The reversion to description that this map represents is only apparent because from the start we are asked to see it "under the theoretical point of view of the last chapter" (123). Given this point of view, the map is at the same time a representation of the network of Pacific reef-island systems and of a theory of their distribution in space and time; it is "corroborative of the truth of the theory" that reef-island

6. From the southern end of the Low Archipelago (the dark mass at the lower margin) to the northern end of the Marshall Archipelago (the dark mass at the fold) the distance is 4,500 miles, the length of a round trip between New York City and El Paso.

systems are distributed according to whether the surrounding areas have subsided, been elevated, or remained stationary (124).

Darwin's colored map is at the same time iconic (representing the spatial relationships and shapes of atolls, barriers, reefs, and large islands in the Pacific) and indexical (it is an effect that points to its cause). Finally, the map is symbolic—it now stands for the truth of a new theory.

Darwin's theory, while not without some problems as to detail, has proven remarkably robust over time: "Previous discussions of coral and reefs existed, but Darwin's was the first and arguably the last successful attempt at a global synthesis" (Rosen 1982, 519). Indeed, Darwin's insight into the calculable nexus of forces behind the character and distribution of reef-island systems looks forward to two later theories with which we will presently deal. One is Wegener's claim that shifts in land-mass positions over time cause major reconfigurations of the earth's surface; the other is that these shifts have seafloor spreading as their motive force.

Figure 5.9 is an embodiment of the latter claim, one consistent with Darwin's theory: in it barrier reefs (solid circles) and atolls (solid triangles) exist predominately in midplate regions, far from the destructive plate boundaries of subduction zones, regions where portions of the earth's tectonic plates dive beneath other plates into the earth's interior. In effect, both of these later

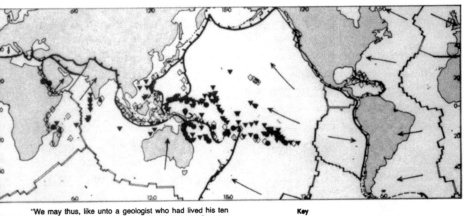

"We may thus, like unto a geologist who had lived his ten thousand years and kept a record of the passing changes, gain some insight into the great system by which the surface of this globe has been broken up, and the land and water interchanged" (Darwin 1845, p. 482). Lithospheric plate boundaries superimposed on Darwin's reef map 1842, less active volcanoes) show how his own views on global geology as supported by reef and volcano evidence (see quotation) fit with the modern plate tectonic view. Note predominance of subsidence reefs in mid-plate regions. For a modern version of Darwin's map, see Schuhmacher 1976.

Key

⌐⌐⌐ Subduction zones (destructive plate boundaries)
——— Transform and strike slip boundaries
——— Mid-ocean ridges (constructive plate boundaries)
- - - - Uncertain boundaries
———→ Directions of plate motions
○ Fringing reefs (new, stable, or elevated reefs)
● Barrier reefs (subsidence reefs)
▼ Atolls (subsidence reefs)

Figure 5.9. Tectonic plate boundaries superimposed on Darwin's map (Rosen 1982, 520). Bora Bora (not visible in map) is at the center in the Pacific tectonic plate, far from the "Ring of Fire," the string of active volcanoes marked by destructive plate boundaries.

theories complete a triumvirate of tectonic forces that account for the surface features of the earth: subsidence, elevation, and horizontal movement.

WEGENER: ON THE ORIGIN OF CONTINENTS AND OCEANS

Although others had suggested continental drift as a cause of the current configurations of the earth's surface, it was Wegener's *Origin of Continents and Oceans* (1st ed., 1915; 4th ed., 1929)[7] that created a lively and long-lasting conversation among the community of geologists as to its truth (Waterschoot van der Gracht 1928; Schwarzbach 1986; Frankel 1987; LeGrand 1988; Oreskes 1999). Throughout his landmark book, by means of words and images, Wegener attempts to convince his readers of a startlingly counterintuitive claim: the earth is "not a solid body . . . but exhibits flow and is subject to continental movements, the wandering of the earth's crust, and probably also the displacement of its axis" (164, 170; translation modified slightly); he attempts to convince us that its continents and islands, once joined into one massive body of land, Pangaea, drifted apart over many eons "like pieces of a cracked ice floe in water" (17). His is an intellectual journey that begins with a tantalizing comment in a 1911 letter to his bride-to-be, Else Köppern: "Doesn't the east coast of South America fit exactly against the west coast of Africa, as if they had once been joined? The fit is even better if you look at a map of the floor of the Atlantic and compare the edges of the drop-off into the ocean basin rather than the current edges of the continents. This is an idea I'll have to pursue" (quoted in Schwarzbach 1986, 76).

The structure of Wegener's provocative argument is inductive. To make his new story of the earth plausible, he relies on a confluence of inductive inferences drawn from very different classes of evidence: "The determination and proof of relative continental displacements," he asserts, "have proceeded purely empirically, that is, by means of the totality of geodetic, geophysical, geological, biological, and paleoclimatic data . . . This is the inductive method" (167). This confluence among classes of evidence transforms Wegener's arguments into what the English philosopher, William Whewell, calls a "consilience of inductions."

Crucial to the persuasiveness of such arguments is surprise experienced at the perception of this hitherto unexpected convergence. Of the Sunda Islands, Wegener asserts that "this region coincides astonishingly [*überra-*

---

7. We use the 1966 English translation of the fourth edition, which originally appeared in 1929 as *Die Entstehung der Kontinente und Ozeane*. In those passages where the German is crucial, the second page number is to the 1929 German edition.

schend] closely with the one where, according to drift theory, the sial [upper layer of the earth's crust] must have been thickened by compression" (91, 93). He indicates further that polar wandering and continental drift "form, in mutual supplementation, the cardinal principle: by its use the previous confusion of disordered, apparently self-contradictory facts links up to form a pattern of a simplicity that astonishes one again and again [*von immer wieder überraschender Einfachheit gliedert*] and is extremely persuasive by virtue of its complete analogy with the present-day climatic system" (145, 151; for additional references, see also 34, 66, 114, 116, 126, 192, 201, 208 in the English edition).

While there is some discomfort among philosophers concerning the incorporation of surprise, a psychological effect, into any theory of confirmation, neither Whewell, the theorist of consilience, nor Wegener, one of its practitioners, has any scruples in so doing (compare Laudan 1971; Fish 1985; Forster 1988). This is because the effect is not merely psychological. If the theory that the continents are fixed in place is presupposed, the converging evidence for drift is bound to be surprising; moreover, this surprise is only the psychological correlative of an underlying epistemological phenomenon: the avalanche of converging evidence that decisively undermines the received theory and at the same time supports the competitor Wegener champions. The degree of surprise incident on consilience, then, corresponds roughly to the degree to which the received theory of a stable earth, the theory under attack, fails.

Because Wegener wishes to make continental drift—a phenomenon that his readers cannot experience—as present for them as possible, it is unsurprising that virtually all of his inductions have a strong visual component. An inductive argument from geophysics, the subject of chapter 4 in *Origin of Continents and Oceans*, exemplifies Wegener's practice. In this chapter, he states a well-known law of geophysics: "that there are two preferential levels for the world's surface which occur in alteration side by side and are represented by the continents and the ocean floors, respectively" (37). The challenge is "to explain this law" (37).

Figure 5.10 exemplifies the law, transforming a host of individual measurements into a hypsometric curve, one in which the earth's distribution of land and ocean floor is plotted as a function of elevation. This graph has been constructed from a base of cross-sectional measurements that subdivides the surface of the earth into one kilometer squares and subsequently arranges this data in order of height above and below sea level (35–36). The key trend is the great difference between the average elevations of the land and ocean floor, a trend made apparent by scanning along the horizontal solid

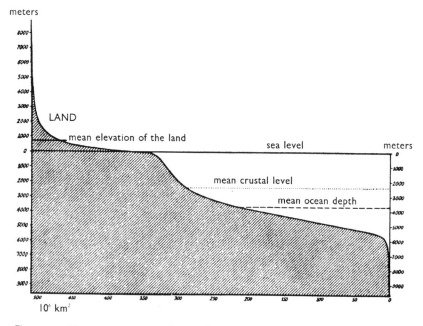

Figure 5.10. Hypsometric curve indicating how much of the earth's land area and mean ocean area is at a given elevation relative to sea level (Wegener 1929, 35).

line labeled "mean elevation of the land" and the dashed line labeled "mean ocean depth," then scanning to and matching with the corresponding numbers on the x- and y-axes: "To put it in rather picturesque terms, the two layers behave like open water and large ice floes" (1929, 37). This coded representation of Cartesian space transforms Wegener's data into a symbolic depiction of the earth's surface, a depiction with indexical implications.

In figure 5.11, Wegener plots a similar set of data so that the trend is visually unequivocal, a solid curve with twin peaks at 100 meters above sea level and 4700 meters below sea level. On this curve Wegener superimposes its normal counterpart, a dotted bell-shaped curve with single peak at about 2,500 meters below sea level. This is the curve that would have been generated had the earth consisted of one uniform layer and not two (Douglas and Douglas 1923). The superimposed curve creates a striking Gestalt contrast. The implication is indexical, pointing to a calculable nexus of forces: because bimodal distribution is real, continental drift may be real. As a consequence of its transformation from the iconic to the symbolic, from data to its geometric representation, figure 5.11 has become indexically transparent. This transparency is in accord with Bertin's insight that "the most efficient constructions are those in which any question, whatever its type and level, can be answered in a single instant of perception, that is, IN A SINGLE IMAGE" (1983, 146; Bertin's emphasis). Bertin is clearly talking about *Prägnanz*, a conceptual *Prägnanz*. In the case of figure 5.11, we see at a glance the visual-conceptual point of the diagram, the marked contrast between the bimodal and bell-shaped curve.

In figure 5.12, the solid curve is transformed into its mechanical counterpart, a schematization of its indexical implications. The components of figure 5.12 are differentiated by a visual code: the crust is divided into the land, represented by diagonal hatching; and the ocean floor, represented by the dotted area. The broken horizontal hatching represents the ocean. While Gestalt contrast designates the components, their interaction is suggested by Gestalt proximity. Figure 5.12 is a model that stands for any continental edge. This symbolic transformation is consistent equally with the data and with drift theory. To see drift over geological time, we need only to animate the model, that is, convert a spatial representation into a spatiotemporal one. In this animation, the land, the large diagonally striped block to the left, moves to the right *through* the ocean and the ocean floor. This startling counterintuitive journey is possible because "under forces applied over geological time scales, the earth must behave as a fluid . . . [T]he critical point in time where elastic deformations merge into flow phenomena depends precisely on the viscosity coefficient" (55). It is to this implausible scenario of solid passing through solid to which the 1960s geologists who revived continental drift pointed when called upon to explain Wegener's failure to convince.

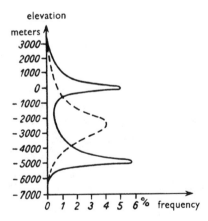

Figure 5.11. Frequency distribution indicating percentage of the earth's land area and mean ocean area that is at a given elevation relative to sea level (Wegener 1929, 35).

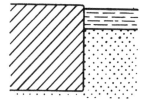

Figure 5.12. Diagrammatic cross-section through a continental margin representing
two-layer model supported by data in figures 5.10 and 5.11 (Wegener 1929, 36). Broken
horizontal hatching = water; dotted area = ocean floor; diagonal hatching = land.

Because geology is a historical science, Wegener's inductive argument
for drift must be transformed into a narrative of drift, a sequence of events
that represents the unique past of the earth. This transformation is foreshadowed early in Wegener's monograph. In the series of maps reproduced
in figure 5.13, Wegener reconstructs the earth's history during three successive geological epochs: the upper carboniferous (about 310 million years
ago), the eocene (around 58 million years ago), and the lower quaternary
(about 1.5 million years ago). As we move from epoch to epoch, the white
space is always in the Gestalt foreground: gradually it emerges as the present configuration of continents and oceans. In addition, the land mass that
was to become South America drifts away from western Africa by the length
of the Atlantic Ocean while, simultaneously, it rotates clockwise to achieve
its current configuration. These three maps—three snapshots in a sequence
that can achieve its full effect only when animated—are not iconic pictures
of the earth; rather, they are conjectures as to its development over geological time. They are symbolic of Wegener's theory. They also embody a narrative that can be responsibly told only after the argument that will support it
has been diligently constructed.

In figures 5.14 and 5.15, the visual hypothesis presented in figure 5.13
is tested against iconic representations of the actual geology and geography
of the continents of Africa and South America and the oceanic rift between
them. The map of western Africa in figure 5.14 depicts "strikes," the intersection of a geological feature and the horizontal plane, rendered visible as
arrays of parallel lines and rendered meaningful as part of a cartographic
code. The direction of these strikes will support a narrative in favor of drift
if, when they are brought together with those in South America, they fit
together, like the pieces of a jigsaw puzzle.

Figure 5.15 is an iconic representation that maps strike directions in
South America, northeasterly and north-south. To see the conformity between the western African strike directions and these, a conformity essential

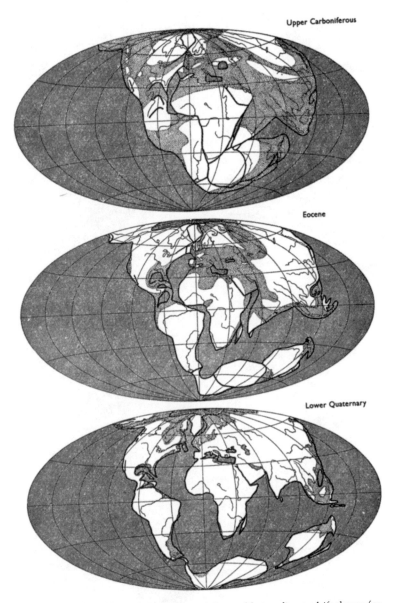

Figure 5.13. Reconstruction of the map of the world according to drift theory for three epochs (Wegener 1929, 18). Present-day outlines and rivers are given simply to aid identification.

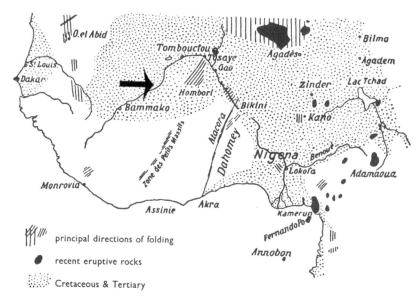

principal directions of folding

recent eruptive rocks

Cretaceous & Tertiary

Figure 5.14. Map of western Africa indicating strike directions (Wegener 1929, 65).

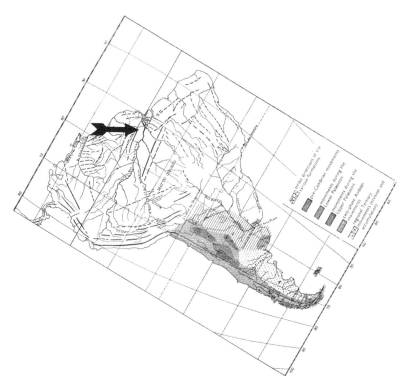

Figure 5.15. Map of South America indicating strike directions (Wegener 1929, 67).

to Wegener's narrative, we have rotated the map counterclockwise in order to take "into account the large angle through which South America must be turned in our reconstruction" (66). As a result of this epistemologically significant rotation, "the direction of the Amazon becomes exactly parallel to that of the upper course of the Niger [the lines just below Tombouctou], so that the two strike directions [along the Amazon] coincide with the African ones" (66; see our two deictic arrows superimposed on figs. 5.14 and 5.15). As a result, the two maps fuse into a single *Prägnanz*, a metonym for the theory of continental drift. The iconic has now been transformed into the indexical, the map into the cause of one of its features, a feature that drift theory has made significant.[8]

Toward the end of his monograph, the transformation from indexical argument into indexical narrative, foreshadowed in figure 5.13, is realized in the fine structure of Wegener's prose. Logical operators, which we have highlighted in italics, simultaneously drive argument and narrative:

> *If* the basalt layer under the granite was really specially fluid as assumed, *then* as the Atlantic rift opened wider progressively, this layer would have had to rise up here, subsequently flowing steadily out from both sides; it would first have formed the whole ocean floor, and would still today form the greater part of it. As the rift opened up progressively wider, the ability of even this material to flow *must finally have* become inadequate, and the underlying dunite *must have been* exposed as windows in the basalt. (211–12)

Wegener's narrative has the dual structure exemplified in this paragraph. On one level, repetitive geophysical processes—including volcanic action, earthquakes, tectonic shifts, elevation, and subsidence—shape the earth's surface; simultaneously, on another level, as a consequence of these repetitive cycles there emerges the unique configuration of the earth's oceans and land masses. Out of Wegener's argument his narrative emerges.

So far we have assumed that there is no discrepancy between what Wegener asserts about the history of the earth and what can legitimately be inferred from the visuals from which his assertions derive; we have assumed that the verbal and the visual are fully compatible. This assumption must now be interrogated. One the one hand, by "synthesizing or explaining a

---

8. We acknowledge that this confluence is visually uncompelling. We will reflect later in this chapter on the contrast between the certainty of Wegener's assertions and the uncertainty of the visual evidence he adduces in their favor.

vast abundance of different sorts of data" (Frankel 1976, 305) from geodetics (astronomical position finding), geology, physics, climatology, paleontology, botany, and zoology, Wegener composed inductive arguments favoring the claim that the continents drifted to their present positions over many eons, as depicted by the chronological narrative implied in figure 5.13. Figures 5.14 and 5.15 and many others in his *Origin* make that case for him. On the other hand, the evidence for his causal argument, embodied in figure 5.12's mechanical model of continental drift, is flimsy: in the opinion of one prominent critic, "having the continents plow their way through the sea floor would be like attempting to thrust a leaden chisel into steel" (Frankel 1987, 210). And for many of his original readers that weak link called into question his entire "drift theory."

Moreover, Wegener's narrative was systematically undermined for many of the geologists in his original audience because, throughout his book, he exhibited a confidence in his theory that his evidence did not warrant.[9] This overconfidence is frequently embodied in Wegener's rhetoric: his was an argument that generated heat as well as light. For example, concerning the continental configuration between South America and Africa displayed in figures 5.14 and 5.15, he said:

> By comparing the geological structure of both sides of the Atlantic we can provide a very clear-cut test [*scharfes Kontrolle*] of our theory that this ocean region is an enormously widened rift whose edges were once directly connected, or so nearly as makes no difference. This is because one would expect that many folds and other formations that arose before the split occurred would conform on both sides, and in fact their terminal sections on either side of the ocean must have been so situated that they appear as direct continuations of each in a reconstruction of the original state of affairs. Since the reconstruction itself is necessarily unambiguous [*zwangsläufige*] because of the well-marked outlines of the margins and allows no scope for juggling [*keinen Spielraum für eine Anpassung*], we have here a totally independent criterion [*einem ganz unabhängigen Kriterium*] of the highest importance [*von größter Bedeutung*] for assessing the correctness [*der Richtigkeit*] of drift theory. (61, 61; same page in both English and German)

---

9. Reading the fourth edition of his book we were left with the impression that a small army of scientists from many specialties held positions consistent with drift theory, but we know from historians such as Frankel (1976, 1987) and Oreskes (1999) that that was not really the case.

This verbal swagger, of which many additional examples may readily be adduced, transforms Wegener from a scientist into an advocate.[10] An advocate's stance leads him systematically to slight his opponents' arguments by means of dismissals so curt as to foster skepticism as to their motive: "these objections, so far as they are not just misunderstandings," he asserts, "mostly involve mere side-issues whose solution would have little significance for the basic concepts of drift theory" (96). At times Wegener goes further; he ignores his opponents' arguments altogether: "We shall refrain here from citing the literature in support of our statements. The obvious needs no backing by outside opinion, and the willfully blind cannot be helped by any means" (133; see also 97, 213). This diminution of his opponents' presence in his text can be visual as well as verbal. Wegener sweeps these opponents under the rug: he buries opposing arguments and their refutations in footnotes (99n; see also 100n, 104n, 113n, 120n, 210n, 215–26n). His treatment of Ökland is a dramatic instance of such interment. In the text, Ökland's views seem in accord with drift theory. Only from the footnote do we learn that he actually rejects drift theory in favor of sunken-continent theory, ignoring, so Wegener asserts, its "geophysical untenability [*Unhaltbarkeit*]" (103n, 124n). In this case, Gestalt foreground and background have been flagrantly manipulated in the interest of advocacy.

Wegener's monograph was a modest success. It was translated into French and English, and his theory was adopted by a few geologists, for the most part those in continental Europe. But its reception in England, and especially in the United States, was largely negative, even hostile (LeGrand 1988, 55–69). The most plausible explanation to date for this relative failure is that of historian Naomi Oreskes (1999). She traces the early rejection of continental drift, especially by the American geological community, to

10. This hyperbolic language is typical. See also *Origin of Continents and Oceans* (Wegener 1929), 60 ("agrees excellently"); 76 ("an almost incontrovertible proof"); 77 ("perfect contact"); 84 ("completely confirmed"); 85 ("completely . . . agree"); 91 ("coincides astonishingly closely"); 95 ("impossible to escape"); 121–22 ("the most remarkable evidence for that fact that drift theory is inescapable"); 126 ("shows surprisingly"); 127 ("inescapably confirmed"); 136 ("No better corroboration of our theory could be desired"); 138 ("this conclusion is quite inescapable"); ("*The conclusion . . . is therefore unavoidable*"); 139 ("The evidence is so compelling"); 142 ("an *utterly compelling reason*"); 145 ("With the present-day position of the continents, however, it is altogether impossible to combine the data into an intelligible system of climates"); 168 ("The westward drift of the continental blocks is immediately evident"); 178 ("This interpretation is rendered specially probable"); 194 ("There is no doubt that all these things are causally connected with one another"); 203 ("There are two factors which corroborate this concept"); 217 ("While this book was in proof, a confirmation of the increase in distance between North America and Europe, claimed in Chapter 3, has been provided"). [Italics in the original.]

two interlinked preferences. The first is a methodological preference for evidence that appears to be independent because its source, instrumental readings, has apparently eliminated the subjective judgment of the observer (Oreskes 1999, 304). The second is the identification of false but useful theories as the truth. The belief that the continents and oceans were always more or less as they are now—"was *enabling*. It enabled [American geologists] to interpret field evidence in a consistent and logical way" (314; her emphasis).

To these explanations of the American and English resistance to drift, we would like to add two. The first is the sheer geophysical implausibility of Wegener's theory: the vision of the essentially solid moving through the essentially solid. The second is his systematic violation of the norms of scientific argument, exhibited most dramatically in the contrast between what his visuals equivocally suggest and what his words unequivocally assert. For example, the conformity between figures 5.14 and 5.15 is hardly compelling. To say this is not to endorse the view that Wegener's advocacy had no impact on the history of geology. On the contrary, a small band of geologists saw his stridency as no more than a sign of frustration, the rage of a distinguished meteorologist whose insight the geological establishment had refused to take seriously.

SEAFLOOR SPREADING AND THE ORIGIN OF CONTINENTS AND OCEANS

Due to the effects of this small band of drift advocates, just after the middle of the century, the fortunes of this theory changed dramatically. These geologists began to gather geomagnetic evidence from rocks around the globe; they also began to gather such evidence from the seafloor. Taken together, their endeavors suggested a new view of the history of the earth: the spreading of the seafloor, now being tracked by the geomagnetic patterns it created, offered for the first time a plausible mechanism for continental drift. Moreover, the timing of this magmatic spread, firmly established by paleomagnetic measurements, constituted a reliable geochronometer of the progress of drift in which "a continent moves *with* the adjacent ocean floor, and not *through* it" (Cox 1973, 15).

The question of whether this new view of an old theory was revolutionary, whether it went on to constitute a new "paradigm," we leave to philosophers. The evidence is overwhelming, however, that it was perceived by participants as the best of professional gifts, an extended horizon of inter-

esting problems without which distinguished careers cannot be made. Jack
Oliver, one of the coauthors of the instant classic "Seismology and the New
Global Tectonics" (Isacks et al. 1968), expresses this well:

> Our paper did indeed support the hypothesis and it emphasized, almost
> to the exclusion of the other versions, the moving plate model. It was
> very exciting to write that paper for we sensed that we were involved in
> a major upheaval in the earth sciences in general, and in seismology in
> particular. Even if time should prove the entire concept in error, I am
> sure I would continue to think of those years as highlights of my scien-
> tific career, for surely the events transpiring then marked the start of a
> new era in the study of the earth. (Cox 1973, 291)

It is to the establishment of the tectonic model based on seafloor spreading
that we now turn. We shall see that this establishment hinged crucially on
the epistemic role of images.

To demonstrate the limits of a proposition-based approach to scientific
knowledge, and to illustrate the strengths of his rival model-based approach,
philosopher of science Ronald Giere (1996) needed a dramatic example that
favored his view. His choice was the decision at the April 1966 meeting
of the American Geophysical Society to support seafloor spreading as the
cause of continental drift. In his view, this choice of a mobilist over a stabilist earth was a matter of fit, specifically the fit among three images of paleomagnetism derived from three different sources: "scattered continental lava
flows, deep sea sediments, and the floor of several different oceans" (299).
Once geologists actually saw this fit, Giere maintains, their conclusion was
inevitable that a mobilist earth must be preferred. In the face of a coincidence of patterns so blatant, a stable earth could not be plausibly imagined.
But no search of the literature will reveal a depiction of this tri-fold fit; only
when these three images of paleomagnetism were shown together in the
April 1966 meeting did geologists conclude, in the words of geologist Allan
Cox, that "there was just no question any more that the sea-floor spreading
idea was right" (quoted in Giere 1996, 299). In effect, at this meeting these
depictions of data trends achieved transcendental symbolic status: they
stood for all the images that had been gathered so far. Simultaneously, they
underwent another transformation: they achieved indexical status, pointing unerringly to a plausible mechanism of continental drift. In Giere's language, the three images were perceived as instances of the same model.

In support of his conclusion that visual coincidence was important in solving this geologic problem, Giere does not mention the reminiscences of several geologists whose work contributed to the creation of this new solution. These strongly support his conclusion. James Heirtzler writes, "I can distinctly remember the night when, for the first time, I could lay 13 trans-Atlantic magnetic profiles on the top of my desk for cross comparison" (quoted in Cox 1973, 228). Tania Atwater captures the transition from speculation to theory, from models of the real world to the real world itself:

> At first Bill [T. W. Menard] and I were catching each other at odd moments, scribbling sketches on envelopes and scraps of paper, but we got more and more excited until we began hunting each other up in the morning to compare the previous night's thoughts. Whenever we found a new geometrical relationship, he could think for a moment and draw out of his mind some appropriate examples from the real world. The creation of brand new fracture zones by changes in direction of spreading was a prediction that fell straight out of the sketching games; there was no well-documented case. We went ahead and published it—his enthusiastic optimism overriding my trepidation. I was utterly amazed when we got some new lines near the great magnetic bight and the pattern was there, just as predicted. That day I was converted from a person playing a game to a believer. (Cox 1973, 410; see also 535)

While we agree with the broad outlines of Giere's analysis, we would modify certain of his claims. Giere attributes Wegener's failure to convince to the fact that his images of continental fit were insufficient for the geological community to choose decisively between a stabilist and a mobilist earth. In support of this argument, he depicts a drift opponent's counterexample, the subcontinent of Australia and the island of New Guinea superimposed onto the Indian Ocean; their fit with neighboring land masses is very good indeed, even though geophysically absurd. But Wegener was not relying solely on coastline conformity, as Giere also notes (281–84). There were a host of conformities, not the least of which was the remarkable instance of mountain ranges stretching conformably across the now-separated continents of Africa and South America. But despite Wegener's host of instances, geologists remained unconvinced because conformities didn't count, except as corroborative evidence. As Oreskes (1999) has written, the geologists of the time had a strong preference for instrumental evidence.

They still do. Oreskes's case seems particularly strong when we consider Harry Hess's 1960 seafloor spreading model. According to this model, through thermal convection, molten mantle material rises from the Earth's interior along both sides of an oceanic ridge and spreads like an erupting volcano, continually forming new seafloor and causing the tectonic plates to spread apart. Giere lumps this model together with Wegener's as offering no decisive choice between rival hypotheses. This is in some sense true, but there is a significant difference. Unlike Wegener's, Hess's model provided the theoretical framework within which a strong argument could be made once the instrumental evidence whose search it motivated became available. In addition, Hess's model gave Wegener's observations genuine epistemic purchase: "These observations were first integrated into the testable hypothesis known as seafloor spreading by Harry Hess" (Phillips 1974, 21). In contrast, Wegener's model did not even warrant testing because "it was physically impossible for a continent to 'sail like a ship' through the sima [the lower layer of the earth's crust]; and nowhere is there any sea floor deformation ascribable to an on-coming continent" (Dietz 1961, 856). Isacks and coauthors reiterate this objection: Wegener's hypothesis "had not received general acceptance, largely because no satisfactory mechanism had been proposed to explain the movement, without substantial change of form, of the continents through the oceanic crust and the upper mantle" (in Cox 1973, 362).

We would also like to offer a caveat to Giere's interpretation of the revelation of the April 1966 meeting of the American Geophysical Society. Figure 5.16 reproduces the published versions of the three visuals presented at the meeting. All three result from the quantitative "study of the natural magnetism of rocks and baked clay [from seafloor samples near ridges] which retain a magnetic memory of the earth's field in the past" (Cox 1969, 237). None of the three is comprehensible without verbal explanation. For that, we draw upon the research articles in which they first appeared.

In the first visual (fig. 5.16, top),[11] we must consult the accompanying text. In it Opdyke et al. (1966, 350) direct our attention to the left-hand column, representing data trends published earlier from lava flow samples. That column represents the earth's magnetic reversals over close to four million years, divided into long time spans at a given polarity ("epochs")

---

11. We ignore as irrelevant to our purpose the geological composition of the deep-sea cores, indicated in the key, the Greek letters indicating zones representative of particular fossil faunal assemblages, the arrow under column V16–133 indicating the boundary above which iced-rafted debris was found, and the plus and minus signs indicating normal or reversed magnetism.

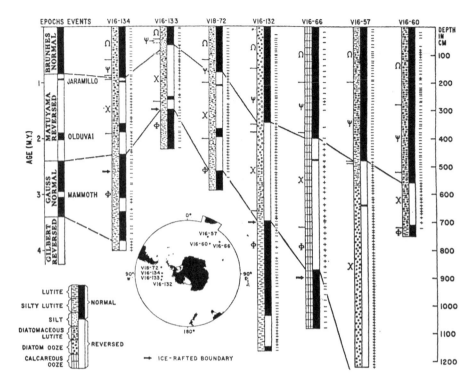

Potassium—argon time scales

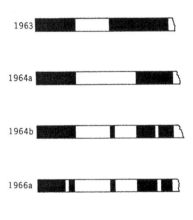

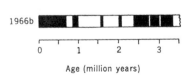

Figure 5.16. Magnetic anomalies in (top) deep-sea cores (Opdyke et al. 1966, 350), (left side) lava flows (Cox 1969, 239), and (next page) ocean basins (Pitman and Heirtzler 1966, 1166). Reprinted with permission from AAAS.

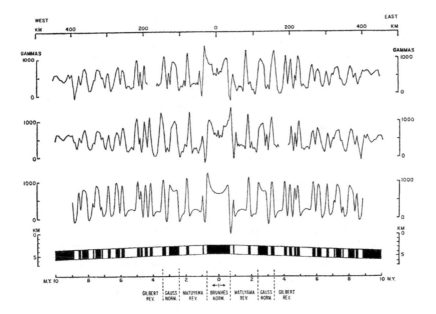

and much briefer periods ("events"). As the authors explain, the alternating
black and white rectangles within a column represent

> the Brunhes normal epoch, o to 0.7 [millions of years]; the Matuyama re-
> versed epoch, 0.7 to 2.4 ± 0.1; the Gauss normal epoch, 2.4 ± 0.1 to 3.35.
> The Gilbert reversed epoch ended at 3.35 million years and its beginning
> is not known. Within the Matuyama reversed epoch, two short periods
> of normal polarity occurred at about 0.9 and 1.9 million years, and these
> have been termed the Jaramillo and the Olduvai events, respectively.
> Within the Gauss normal epoch there was a short period of reversed po-
> larity at 3.0 million years, called the Mammoth event. (349)

From the left-hand column, the dashed lines draw us to the next column
to the right. At the top of the column appears the notation "V16–134,"
the identification number assigned to one of the seven deep-sea cores from
which magnetic measurements were made. Scanning down to the encircled
map projection, we can spot the exact location in Antarctica where this
particular core sample was taken, the basin of the Antarctic Bellingshausen
Sea. Scanning and matching the two adjacent columns linked by dashed
lines, we immediately notice the closeness in the relative lengths of the
black and white vertical bars. We will not discuss the other columns; we
will only remark that all show a similar pattern of magnetic reversals. The

authors' interpretation of this visual is that "some deep-sea cores contain a complete or nearly complete record of the history of the earth's magnetic field back to about 3.5 million years" (356). They do not mention seafloor spreading and its effect on the tectonic plates.

The second visual (fig. 5.16, bottom left) rotates the left-hand column of the top visual ninety degrees and eliminates the distance scale; its intervals represent magnetic reversals and are based on age and magnetic measurements of seafloor volcanic rock. The number to the left of each bar indicates the year the measurements were made. By scanning and matching the five bars from top to bottom, we can perceive the rapid evolution in understanding of the polarity epochs and events. The top two bars (1963 and 1964a) indicate two polarity epochs of nearly equal time scale; the bottom bar (1966b), the irregularly timed epochs and events discussed in the previous paragraph. This visual traces the rapidly changing interpretation of the acquired data on magnetic reversals evident in the seafloor over geologic time.

The third and last visual (fig. 5.16 on p. 193) illustrates the magnetic fluctuations over more than 400 kilometers on either side of a seafloor ridge in an oceanic basin. To make sense of the data trends over time, we must scan and match the three rapidly fluctuating curves with the scale to the left and right, magnetic activity measured in gammas; with the bottom scale, time in millions of years; and with the top scale, distance in kilometers from the ridge. The middle curve is the magnetic fluctuation profile generated from the data. The one immediately above is its mirror image, indicating bilateral symmetry starting from the ridge in the middle (marked as zero). This symmetry is striking, constituting strong evidence for evenly distributed seafloor spreading on both sides of an active ridge. Furthermore, these magnetic fluctuations correlate impressively with the known epochs in the earth's magnetic field, represented by the horizontal black-and-white array at the foot of the diagram. They correlate as well as with the bottom curve just above that array, its magnetic fluctuations derived from a theoretical model. In other words, actual seafloor spreading conforms to the model and known data and is, above and beyond this, almost perfectly symmetrical on either side of its generating ridge.

In Giere's view, the geologists at the April 1966 meeting of the American Geophysical Society inferred that no stabilist earth could have produced this coincidence among these three visuals, especially the remarkable coincidence clearly evident by scanning and matching the sets of black and white rectangles present in each of the three, representing, respectively, periods of normal and reversed magnetism in deep-sea cores, lava flows, and the seafloor. It is true that, because each counts as evidence of worldwide polar magnet reversals, the three visuals share the same cause and, there-fore, are mutually reinforcing. But in our view they are not rhetorically equivalent. To attain the subjective certainty to which Cox attests (Giere 1996, 299), the geologists present need only have focused on the third of the visuals we reproduce, the Eltanin-19 profile, named for the nineteenth pass of the research ship, *Eltanin*, over the Pacific-Antarctic Ridge. The graphi-cal creativity displayed therein is remarkable, bestowing presence on three startling visual coincidences: the symmetry in magnetic field reversals on both sides of the ridge, the coincidence between theory and the Eltanin data, and the coincidence between theory and earlier data on magnetic re-versals. What the geologists present actually did think at the April meet-ing, of course, is lost to history. But in view of its rhetorical superiority, the Eltanin-19 profile in itself provides convincing evidence of seafloor spread-ing, all the more convincing when placed in the context of the evidence of parallel magnetic reversals of the other two visuals. In other words, in our view, the Eltanin-19 profile counted as convincing, the other profiles as cor-roborating evidence. Cox confirms this view; he relates the conversion of geologist Lynn Sykes after Sykes was shown "the amazingly symmetrical Eltanin-19 magnetic profile across the East Pacific Rise. Sykes was not the first skeptic to be converted by that 'magic profile'" (290).

Finally, we would like to extend Giere's analysis to include the narra-tive aspect of geology. An early advocate of seafloor spreading, Hess stated in a 1962 article that if his proposal were accepted, "a rather reasonable story could be constructed to describe the evolution of ocean basins and the waters within them" (in Cox 1973, 27). At the end of his paper, he outlines this compelling and counterintuitive tale:

> The Atlantic, Indian, and Arctic oceans are surrounded by the trailing edges of continents moving away from them, whereas the Pacific Ocean is faced by the leading edges of continents moving toward the island arcs and representing downward flowing limbs of mantle convection cells or, as in the case of the eastern Pacific margin, they have plunged into and in part overridden the zone of strong deformation over the downward-flowing limbs. (37)

Hess avers that, however erroneous in its details, this story "appears to be a useful framework for testing various and sundry groups of hypoth-eses related to the oceans" (38), a speculation that other geologists turned into a narrative reality. In 1968, Heirtzler and his collaborators extrapo-lated a time scale from magnetic anomalies. On the basis of this extrapolation, they refined Hess's narrative. For example, they theorized that "the split in the South American and African continents may have begun in [the] early Mesozoic [around 250 million years ago] with the break away of the African-South American block" (Cox 1973, 281). Two years later, on the basis of the firmer evidence of deep-sea drilling in the South Atlantic, Arthur E. Maxwell and his team revised this estimate: "with the assumption of an average spreading rate of 2 centimeters per year, it is possible to estimate that the separation of South America and Africa began some 130 million years ago, or during the early Cretaceous" (Cox 1973, 579).

Tanya Atwater added to this story, extending these inferences from ocean basins to continents, specifically to the continent of North America. Her 1970 paper has the additional virtue of making it especially clear that these geological arguments proceed at two levels. On one level, she establishes seafloor spreading as a geological reality; on another, she explains the ways in which seafloor spreading shapes the oceans and their adjoining continents. In the northeastern Pacific, for example, she infers from the magnetic anomalies the existence of a tectonic plate that no longer exists, the Farallon plate, detected because the anomalies currently in existence

> represent only the western half of the symmetrical pattern expected. All presently known ridges spread approximately symmetrically, and so we might expect to find the other halves of these anomalies somewhere. In fact, the eastern halves and the ridge itself are for the most part missing. This geometry indicates that that there once was another plate lying to the east of the ridge, the "Farallon plate" of McKensie and Morgan (1969), which contained the anomalies. (Cox 1973, 585)

This chain of inference can now conclude with a story: the Farallon plate, proceeding eastward at 5 centimeters a year, is gradually consumed by a now-absent trench until "as recently as 29 to 24 [million years] ago" (586).

Unlike Wegener, Hess and his fellow advocates of seafloor spreading were careful to define exactly the level of belief to which their claims were entitled. Their reflections belie any notion that scientists are as a rule philosophically naïve. Hess says he considers his paper "an essay in geopoetry" (Cox 1973, 23). He has "attempted to *invent* an evolution of the ocean basins" (38; our emphasis). Writing in 1967, after the general acceptance of seafloor spreading, Dan P. McKenzie and Robert L. Parker say "it is, however, only an instantaneous phenomenological theory" (Cox 1973, 63), one that cannot account for either the evolution of the tectonic plates or the mechanism driving their spreading. Although Atwater regards her conclusion about the Farallon plate as "nearly inescapable," she feels that her other conclusions are less certain. She regards her extrapolation of continental tectonics to the early Cenozoic (about 65 million years ago) as "very tenuous" (605); indeed, in her brief discussion section she labels it "outrageous" (609). James Gilluly also asserts in 1971 that plate tectonics cannot be the whole story: "many tectonic and magmatic features are so situated that any connection with plate tectonics is so tenuous as to be visible only to the eye of faith" (648). Unlike Wegener, these geologists inferred from their images only what was, in fact, inferable according to the consensus of their professional colleagues at the time of publication.

CONCLUSION

This chapter traces the development of geologic narratives from the early nineteenth through the second half of the twentieth century. We follow Charles Darwin as he re-creates the history of the coral reefs, then Alfred Wegener as he reenvisions the drift of the continents. In a final section, we follow the progress of a hypothesized cause of drift, seafloor spreading, as it wends its way from conjecture to accepted fact. The multimodal texts that we have examined take full advantage of the communicative utility of scientific visuals for representing not only space above and below ground, but time—in particular, unimaginably long stretches of past time. By means of words and images their authors construct new stories of the history of the earth—stories whose time frame reaches back as far as hundreds of millions of years.

In Darwin's *The Structure and Distribution of Coral Reefs*, verbal descriptions are transformed by visual evidence into an argument in favor of his geological theory of subsidence, explaining the origin and history of coral reefs in the course of deep time. The text is organized so that it moves deliberately from the verbal description, visual depiction, and analysis of one reef-island system to the verbal description, visual depiction, and analysis of all. Within that narrative structure, Darwin argues that subsidence exists, is competent to cause the evolution of reef-island systems, and is in fact responsible for that evolution. Wegener's *The Origin of Continents and Oceans* also hypothesizes an earth with a new history. Over vast amounts of time, he asserts, the continents have drifted over enormous distances to reach their present positions. His book has a dual structure. On one level, geophysical processes cycle and re-cycle; on another, as in figure 5.13, a plot unfolds from which there emerges the unique configuration of the earth's oceans and land masses.

In contrast to *The Structure and Distribution of Coral Reefs*, Wegener's story that the continents drifted apart over deep time was not backed up by arguments that the majority of the geological community at the time found convincing. In the final edition of Wegener's book, the concerted and spirited rejection of many of his colleagues provoked him to a strident rhetoric that too often substituted for sound argument. Still, Wegener's rhetoric can be seen as more than a hollow gesture born of desperation; it can be seen as a placeholder for evidence not yet available, a sign that a first-rate scientist was willing to stake his reputation on the truth of his mature intuition, even in the face of withering criticism. It is in large part because of Wegener's obstinacy that continental drift remained into the mid-twentieth century a problem open to geologists of the caliber of Allan Cox, James Heirtzler, and Tanya Atwater. And it was through the efforts of this small group of talented believers that the problem of continental drift was, finally, solved. This controversy reached a dramatic climax when, at a scientific meeting, three crucial visuals related to seafloor spreading were presented.

In the following chapter, we broaden the context in which scientific claims are accepted or rejected to include not only science itself, but the culture at large. We consider the Victorian reaction to deep time, the idea that the earth and the creatures that inhabit it, including humankind, evolved over eons whose extent was, at the time, unimaginable. Pictures and "argumentative narratives" also proved crucial in making that case.

CHAPTER SIX

Verbal-Visual Interaction in the Victorian Discovery of Deep Time

It is very remarkable that while the words *Eternal, Eternity, Forever*, are constantly in our mouths, and applied without hesitation, we yet experience considerable difficulty in contemplating any definite term which bears a very large proportion to the brief cycles of our petty chronicles. There are many minds that would not for an instant doubt the God of Nature to have *existed from all Eternity*, and would yet reject as preposterous the idea of going back a million of years in the History of *His Works*. Yet what is a million, or a million million, of solar revolutions to an Eternity?
—George Scrope, *Geology of Central France* (1827)

In this chapter, we explore the interaction of words and images by means of which the Victorian era grappled with a disturbing discovery—the vast amounts of time it took for the earth, for living creatures, and for humankind to evolve. In the first edition of *The Origin of Species* (1859), Darwin provided a quantitative estimate of the vast expanse of geological time needed to "denude" the once heavily forested Weald in southeast England:

it is an admirable lesson to stand on the North Downs and to look at the distant South Downs; for, remembering that at no great distance to the west the northern and southern escarpments meet and close, one can safely picture to oneself the great dome of rocks which must have covered up the Weald within so limited a period since the latter part of the Chalk formation. . . . under ordinary circumstances, I conclude that for a cliff 500 feet in height, a denudation of one inch per century for the whole length would be an ample allowance. At this rate, on the above

data, the denudation of the Weald must have required 306,662,400 years;
or say three hundred million years. (287)

It was a calculation Darwin was soon to regret. In 1862, the most respected
physicist in England, Lord Kelvin, confronted him on this very point. Kelvin
argued that according to the second law of thermodynamics, the law of en-
tropy, the earth could not be as old as geological or biological evolution required. Until the discovery of radioactivity, a more enduring source of heat,
there was no sound counterargument to Kelvin's claim (Burchfield 1990). In
a brace of rhetorical questions, Kelvin reduced the apparently plausible to
the essentially ridiculous:

> What then are we to think of such geological estimates as 300,000,000
> years for the "denudation of the Weald?" Whether is it more probable
> that the physical conditions of the sun's matter differ 1,000 times more
> than dynamics compel us to suppose they differ from those of matter
> in our laboratories, or that a stormy sea, with possibly Channel tides
> of extreme violence, should encroach on a chalk cliff 1,000 times more
> rapidly than Mr. Darwin's estimate of one inch per century? (quoted in
> Burchfield 1990, 32)[1]

Darwin had to take this criticism seriously, especially when Kelvin's friend
and associate Fleeming Jenkin incorporated this argument in a scathing review of Origin: "Not only is the time of the earth limited, but it is limited
to periods utterly inadequate for the production of species according to Darwin's views" (cited in Hull 1973, 327). Darwin removed the offending passage from all subsequent editions of Origin. In his analysis of the problem,
philosopher David Hull underlines the arrogance of physicists in assuming that the findings of their science trumped the findings of the biologists
(349). After all, Darwin's was, as it turned out, a good estimate. But to refute
Kelvin was futile, given his prestige and that of physics. In the latter half of
the nineteenth century, evolutionists and geologists were alike faced with
a formidable intellectual obstacle: deep time was, by far, not deep enough.
   While only geologists and evolutionists were confronted with Kelvin's
argument, those interested in the descent of humankind joined their fellow
scientists in facing an equally formidable obstacle. The advocacy of deep

---

   1. Note that the 300 million years estimated by Darwin only covers the Weald erosion, not
its original formation and development. Formation and erosion to its current state would have
required billions of years (Lewis 2000, 25).

time had to make its way despite an uneasiness deeply rooted in the Judeo-Christian tradition, a set of intellectual and emotional habits of mind according to which human beings are central; and human time, the only time. The biblical universe is temporally and spatially compact: within a span of about two hundred generations, the earth, comfortably surrounded by its sun, its moon, and its stars, forms a stage on which human beings play their leading roles. Nor is God indifferent to human fate; indeed, the human and the divine are intimately involved. The world of the New Testament differs from the Old only in the sense that the temporal and the eternal are fully synchronized. What we do here registers above; our eternal fate depends on our time on Earth. Dante's *Divine Comedy* is the supreme expression of a cosmological view that permitted Bishop Ussher to set 4004 BCE as the firm date of creation. By the time Victoria ascended to the throne in 1837, although the universe had expanded spatially for many scientists and for the vast majority of the public, humankind was still at its center. Moreover, temporally, it remained essentially the same universe: time was still human time.

How would it be possible to overcome the disenchantment this new view entailed, in which time expanded while human importance shrank? How would it be possible to avert the danger of alienating human beings from their own past? To counteract this potentially debilitating effect, natural philosophers who focused on geology, evolution, and anthropology needed to turn the earth into *our* earth, living things into *our* relatives, and early hominids into members of *our* human family. As a consequence of this imaginative transformation, we would retain our centrality: like Adam, we would give every creature its proper name. Human beings would remain central in another sense. If of necessity deep time stripped the biblical view of many of its more familiar trappings, it might still be possible to maintain one of its essential features, to discover that human beings were still the pinnacle of creation.

The present chapter concerns the efforts of nineteenth-century geology, evolutionary biology, and anthropology to domesticate the notion of deep time for a Victorian audience. The popularity of this topic in nineteenth-century Britain is hard to exaggerate. The audience included not only those actively engaged in research in the historical sciences, but also hordes of science enthusiasts, both men and women: "mechanized presses, machine-made paper, railway distribution, improved education, and the penny post played a major part in opening the floodgates to a vastly increased reading public" (Secord 2000, 5). Relatively large numbers of readers bought and digested lengthy tomes that sought radically to alter reigning notions about the life of the earth and life on earth. One of the first books on the subject appeared in the years 1830–1833, the three volumes of Charles Lyell's *Principles of Geology*. It went through eleven editions in Lyell's lifetime. The *Origin of Species*, first published in 1859, went through six editions in Darwin's lifetime. A few years after the debut of *Origin*, in 1863, Lyell published *The Geological Evidences of the Antiquity of Man*. In this book, he presented extensive evidence that for a long period much of the earth was covered by a sheet of ice perhaps a mile thick. Following this Ice Age, human beings inhabited the earth from a time far earlier than biblical chronology would sanction. William Whewell wrote to the author that he was reading *Geological Evidences* "as all the world is doing or has done" (Grayson 1983, 200). At least initially, Lyell's work outperformed Darwin's in the literary marketplace. Within a week of its publication, the four thousand copies of the first edition had been sold (Cohen 1998, 90). Two more editions appeared within the next year. Lyell significantly revised the book in a fourth edition published in 1873.

In this chapter, we investigate the interaction of words and images in four representative and profusely illustrated works from that era: James Geikie's *The Great Ice Age and Its Relation to the Antiquity of Man* (1894), Charles Darwin's *The Descent of Man, and Selection in Relation to Sex* (1871) and his *The Expression of the Emotions in Man and Animals* (1872), and John Lubbock's *Pre-Historic Times, as Illustrated by Ancient Remains, and the Manners and Customs of Modern Savages* (1865). Most of the images are realistic drawings of the natural world not dissimilar to Perrault's seventeenth-century drawing of a chameleon (fig. 3.10); the difference is that Geikie, Darwin, and Lubbock interpret their images as the products of geological and biological forces operating over deep time. Animated by their verbal contexts, these images contribute to a view of a past vastly different from the present, one shaped and framed by a calculable nexus of forces. Without the surrounding texts, of course, such images are nothing more than pictures of natural and man-made objects. When we look at an image of a flint, unless we are experts, we cannot tell whether its shape is due to human or natural forces, and, if human, we cannot, unless we are experts of a different order, tell whether it is a tool or a weapon, or whether it was made yesterday or a million years ago. It is we—or rather the natural philosophers we deputize—who employ words to turn these images into icons of a past age, symbols for the vastness of that past, and indices of the operation of past practices and geological and evolutionary forces. It is these semiotic transformations, these instances of verbal-visual interaction, that we now investigate.

GEOLOGICAL DEEP TIME

In the decades after Lyell's *Geological Evidences* appeared, a raft of books on the history of the earth and its inhabitants also appeared,[2] among them *The Great Ice Age and Its Relation to the Antiquity of Man* by James Geikie, three editions of which were published in his lifetime (in 1874, 1877, and 1894). It is on this work that we now focus. Geikie's objective was to convince his readers that our prehistoric ancestors viewed a landscape that had been shaped by eons of geological action, a landscape still in the process of formation:

> We are introduced to scenes that are in strangest contrast to what now meets the eye in these latitudes [Great Britain and Europe]: changes of the most stupendous character pass before us; we see our islands and northern Europe at one time enveloped in snow and ice, while from the Alps and other mountain areas enormous glaciers descend to the low grounds. At another time the British Islands are united to the Continent, and the land is occupied by savage men and animals, many of which have long since vanished from the European fauna. In a word, epochs of arctic severity are beheld alternating with epochs of genial climatic conditions—each accompanied by less or greater geographical changes— throughout a protracted cycle of time. (1894, 2)

Immediately to involve his readers as partners in the investigation of this past, Geikie adopts a strategy not dissimilar to that of Edgar Allen Poe's "The Purloined Letter." In this famous detective story, C. Auguste Dupin discovers the letter in question hidden in plain sight; analogously, Geikie permits his readers to discover that a careful scrutiny of the current landscape is the key to the decipherment of our geological past. Like the purloined letter, the clues to the earth's deep past are hidden in plain sight, though only with the guidance of the knowledgeable natural philosopher can they be discovered and deciphered. Figure 6.1 illustrates this process. It is a drawing by Benjamin Peach, the great Scottish geologist and a fine draftsman. In the then-absence of a convenient means of reproducing

2. Among prominent such books we would also include John Lubbock's *Pre-Historic Times* (1865); John Evans's *Ancient Stone Implements, Weapons, and Ornaments of Great Britain* (1872); Hodder M. Westropp's *Pre-Historic Phases* (1872); James Geikie's *Prehistoric Europe: A Geological Sketch* (1881); W. Boyd Dawkins's *Cave Hunting* (1874) and his *Early Man in Britain* (1880); and Charles Darwin's *The Descent of Man, and Selection in Relation to Sex* (1871) and his *The Expression of the Emotions in Man and Animals* (1872).

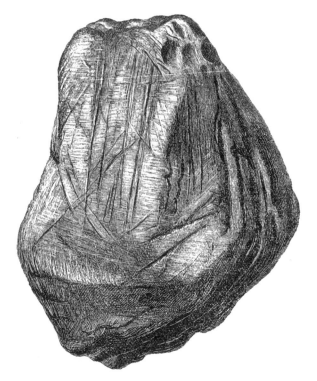

Figure 6.1. Realistic drawing of scratched stone of black shale (typically several feet to yards in diameter) from the till (Geikie 1895, 8).

photographs in mass-produced books, this drawing is its epistemological equivalent: it is meant to count as an accurate representation of what readers might see on a Sunday's stroll in the mountains. The presupposition of accurate representation is vital since, as we shall see, this stone will count as evidence in a scientific argument.

Two of its features are immediately apparent: its generally smooth surface, and the more or less regimented scratches on that surface, standing in the foreground. These, which contrast in depth and breadth, form two clusters. The main cluster is north-south in parallel lines; the other is oriented toward the northeast to southwest, though in a less regular fashion. According to Geikie, "no other appearances connected with [the geological feature known as] the till are more striking than these scratchings and smoothings. They become to the geologist what hieroglyphics are to the Egyptologist—the silent but impressive records of an age long passed away, enabling him to realize the former existence in these islands of a state of things very different indeed to that which now obtains" (12). Because these scratchings constitute evidence for deep time, they acquire a name: striae. It is this verbal baptism that confers geological status on this stone, that turns the iconic into the indexical, pointing to a cause.

Fifty pages later the mystery of the origin of these straie is solved: they have been produced by the action of enormous glaciers moving slowly and steadily over the landscape. The directions of the striae we have just pointed out are coincident with directions in the ice flow:

> Each boulder-clay stone gives evidence of having been subjected to a grinding process. Every fragment has either been jammed into the bottom of a glacier and, held firmly in that position, has been grated along the rocky surface underneath, or over a pavement of the tough stony clay itself; or, enclosed in the slowly moving subglacial debris of gravel, sand, and mud, it has in like manner been brought again and again into forcible contact with the rocky floor, as the material gathering below was pressed and squeezed and rolled forward by the ice. In such a position the stones would naturally arrange themselves in the line of least resistance; hence it is that the most distinct ruts and striae coincide with the longer diameter of the stones. But when the stones and boulders which are dragged on underneath a glacier approach a round or oval shape they can have no tendency to lie in a particular way and so will come to be scratched equally well in all directions. (63)

The semiotic transformation from the iconic to the indexical, from the scratched stones to the scratches as evidence for the passage of glacial ice, is instantiated in this passage's predicates by the careful modulation of tense, voice, aspect, and modality. Such modulations track a semiotic transformation from the iconic to the symbolic and the indexical, from the particular stone depicted in figure 6.1 to all such stones as evidence of gigantic ice floes operating according to natural laws and over extended landscapes over vast stretches of time. In the first sentence, "gives" is in the "eternal" present, the tense of law-like statements, while the string of present perfects ("has . . . been jammed . . . held . . . grated . . . brought") in the next sentence embodies the result over time of the law-like effect the first sentence enunciates. These second-sentence verbs are appropriately in the passive voice: glacial forces are the agent in all cases. By means of the modal "would," the "would" of characteristic activity, the verbs in the third sentence convey the natural necessity imposed by the physical forces mentioned in the first sentence, while a return to the present tense, to "is" and to "coincide," brings readers back to their present. In the fourth sentence, in contrast, the

historical present ("are dragged") projects this same reader into a hypotheti-
cal past as an observer of the glacial forces in action; at the same time, the
modals "can" and "will,"—the predictive "will"—represent the natural necessity those forces impose (Quirk et al. 1972, 61–122).

The same forces that scratch the stones reshape the landscape as a
whole. Geikie begins his analysis of these larger changes with the experience of an average Englishman on a Sunday jaunt: "To the wanderer along
the course of some lowland streams nothing can be more striking than the
sudden and complete change of scenery that ensues upon the passage of a
stream from its new into its old channel" (Geikie 1894, 110). Glacial action
has altered the landscape and, consequently, the direction in which water
can most readily flow. Figure 6.2 is a representation of the effect of these
forces. To clarify its meaning, Geikie employs a visual code of contrasting
symbols. In the lower diagram, the dotted lines indicate the buried path of
the old course of the river, about as wide between its banks as the new. The
arrows represent the direction of flow. Moreover, "from *f* to *f* it will be observed that the present channel coincides exactly with the old course, while
at x x the latter is cut across nearly at right angles by the former" (111). To
get a full picture of the differences in configuration between the new and
old beds, however, an aerial view is insufficient: we need the vertical view
that a section provides, the view of the upper two diagrams. These depictions also indicate that the bed where the older and newer rivers coincided
(upper left diagram) is shallower, the result of "the long-continued action of
springs and frost" (112).

The map in figure 6.3 is a realization of the model depicted in figure 6.2.
At one time, the Tweed flowed not in its present course, but along a course
from Cademuir to Bonnington:

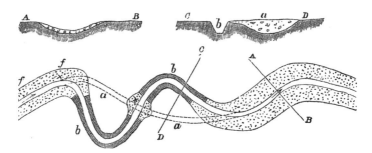

Figure 6.2. Map showing river course during pre- and post-glacial periods (*a*, marks the
buried river course; *b* marks the post-glacial channel). Upper diagrams show valley
along lines *A-B* and *C-D* in lower diagram (Geikie 1894, 111).

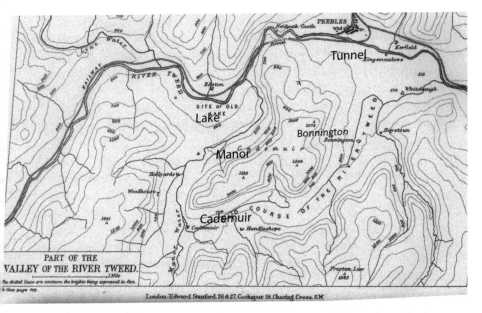

Figure 6.3. Map of the valley of the Tweed, showing the old lake that was the result of the damming up of the Tweed (Geikie 1894, facing 119). Large print labels "Tweed," "Tunnel," "Lake," "Manor," Cademuir," and "Bonnington" have been superimposed.

It will be seen that from Cademuir farm to Bonnington there extends a broad flat hollow, the bottom of which at its highest level is only some hundred feet or so above the level of the Tweed at Neidpath [marked Tunnel on the map]. Were the narrow glen at this latter place to be filled up, the Tweed would be dammed back, a lake would have formed, and thereafter both the Tweed and its affluent, the Manor Water, would flow round by the Cademuir hollow. That such was actually the course of one or both of these streams at a comparatively recent geological date is proved by the fact that the Cademuir hollow is paved with river gravel, which could have come from no other source. (119)

The predicates in this passage move from the present all can experience to the conjectural past expressed in the subjunctive. Finally, the evidence of the river gravel transforms conjecture into fact: the simple past supervenes along with the predictive "could" (Quirk et al. 1972, 61–122).

The surface of the earth told only part of the story; the remaining clues to the mystery of long-term geological change apparently awaited extensive—and prohibitively expensive—excavation. But the problem was soon solved: what geology could not afford to reveal by design, the indus-

trial revolution revealed as its by-product. The building of railroads and the
quarrying of stone that created England's future simultaneously uncovered
its deep past. Figures 6.4 and 6.5 illustrate this process. In the 30-foot section of the quarry depicted in figure 6.4, we see the Hessle clay, h, the product of glaciations, covered by a thin layer of soil. Below the Hessle clay lie
c, the beds that contain numerous fossilized shells of *Corbicula fluminalis*,
a species of freshwater clam. These are interrupted by piles of talus: masses
of angular, jagged rocks.

The most interesting geological feature in this section is the apparently
isolated fragment of Hessle clay, marked b^1 and visible along the vertical
line A-B. Figure 6.5, a slice of the landscape along line A-B in figure 6.4,
makes this presence clear. The diagram is not iconic but symbolic: it stands
for theory. Unlike figure 6.4, it embodies an inference that the fragment b^1
is linked to h, and that c is an intruded tongue of the larger deposit, forced
into its present position by the downward and forward movement of the
glacier.

The most interesting feature of the section, however, is not geological.
The presence of the fossil *Corbicula* in the bed below the glacial deposit
allows us to make a bold retrodiction concerning climatic change during a
succession of glacial eras:

> The very presence of the fresh water shell *Corbicula fluminalis* in the
> Kelsea gravels is suggestive of great geographical and climatic changes.
> This mollusk could not have lived in any English river during the forma-
> tion of the lower glacial deposits. It immigrated into the Humber valley
> after the disappearance of the great ice sheet. Whence did it come? So far
> as I know, the shell occurs only in Pleistocene deposits in the east of En-
> gland, and its peculiar distribution seems to suggest derivation from one
> and the same source. (Geikie 1894, 365–66)

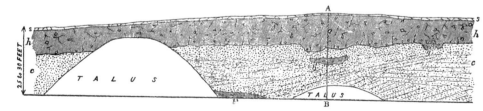

Figure 6.4. Map of geologic section for a large ballast pit (Kelsea Hill, June 1876) to a
depth of 30 feet: s, surface soil; h, Hessle boulder-clay; c, *Cyrena* beds; b^1, intruded mass
of Hessle boulder-clay; g, mingling of beds h and c; b^2, boulder-clay, probably connected
with h; vertical line A-B, slice of diagram section in figure 6.5 (Geikie 1894, 361).

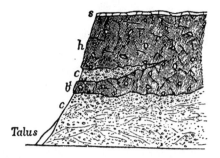

Figure 6.5. Map of geologic section across vertical *A-B* in Figure 6.4: *h*, Hessle boulder-clay; *c, Cyrena* beds; *b¹*, intruded tongue of *h* (Geikie 1894, 362).

What justifies the unqualified assertion of such conjectures? It is Geikie's actualism, his belief that the forces that account for the condition of the earth far back in deep time are essentially the same as those currently operating. To account for the presence of stones and boulders in glacial areas apparently unaffected by glacial action, he cites a remarkable incident of a recovered knapsack:

> The generally non-glaciated aspect of the boulders in question is also perfectly accounted for on the theory I have ventured to advance. Once embedded in the ice, stones and boulders might travel for hundreds of miles without suffering abrasion. The well-known incident of the knapsack which was lost in a crevasse of the Glacier du Talefre on 29th July, 1836, and disgorged by the coalescent Glacier du Lechaud on 24th July, 1846, after having travelled, embedded in the ice, over a distance of 4,300 feet, shows how little change a hard lump of rock would sustain in travelling through a mass of glacier-ice. Occasionally, however, such an included block might be rubbed against the rocks of a hill-side, and so receive a dressing on one or more faces. (204)

Actualism is a presupposition realized in the tenses of his verbs. Early in the book, Geikie employs the present tense to place us in the position of fictional observers of the geological forces in action:[3]

> During the night, and at early morn, dead silence reigns among the snowy peaks: no streams are heard, no water trickles over the surface of the ice; but when the power of the sun begins to be felt, then the noise of

3. See also first Geikie quote (1894, 2) in this section.

water running, leaping, and falling grows upon the ear; soon the glaciers are washed by numberless little streams; great avalanches, wreathed in snow-smoke, rush downwards with a roar like thunder; masses of rock wedged out by the frost of the previous night are now loosened by the sun, and dash headlong down the precipices, while the long trains of débris hurry after them, and are scattered far and wide in the wild confusion along the flanks of the glaciers. (41)

In this passage, each instance of the present tense vacillates productively between what is observed "now" in the historical present and what has always been observable in the eternal present of processes continuously in operation (Quirk et al. 1972, 61–122).

Actualism is also at work in Geikie's visuals. Complementing the above passage is an engraving of an alpine glacier. By itself, this image simply depicts a static mountainous glacier in black and white. But by means of verbal-visual interaction Geikie transforms this image from the present to the eternal present, and from an isolated moment in time to a dynamic natural process. Geikie's visuals also domesticate deep time in another way. His book includes several maps that superimpose the imagined space once occupied by enormous glaciers in the distant past onto the actual space occupied by large land masses in the present (plates facing pp. 437, 465, 475, 490, 691, 724, 727).[4] Our earth becomes at the same time ours and the very different earth of the Ice Age.

In *The Great Ice Age*, deep time has been domesticated: it is now *our* time on *our* earth. The deep past no longer extends beyond our intellectual, emotional, and cultural grasp; instead, it surrounds us as it had surrounded our prehistoric ancestors. *They* have been a witness to it; *we* are *still* a witness to it. It is thus that Geikie created a world in which deep time could be acceptable to Victorian consciousness and conscience; it is thus that he created a scientific and a cultural context in which geological claims that presuppose deep time could be safely made.

EVOLUTIONARY DEEP TIME

In *The Descent of Man and Selection in Relation to Sex* (1871), Darwin creates a world in which the evolution of humankind is possible, a world in which deep time must be presupposed. In pursuit of this goal, he employs

---

4. We discussed another such spatiotemporal map, taken from Lyell's *Antiquity of Man*, in chapter 3 (fig. 3.6).

three argumentative strategies. In the first, he amasses evidence for the gradual emergence of humankind from creatures lower in the evolutionary scale, linked to us by various analogical relationships. Despite these relationships, the gap is huge between us and even our closest living precursors; clearly, then, our gradual evolution to our present state requires deep time for its enactment, an inference that readers may make, but which Darwin scrupulously avoids. Pursuing a second strategy, Darwin links humankind to other creatures by means of a pervasive anthropomorphism. These creatures must be our relatives, he implies, because they behave and think *almost* as we do. To move from *almost* to us, we may infer, though Darwin avoids the inference, requires deep time for its enactment. In a third strategy, Darwin infers evolutionary descent through deep time from a comparison among extant creatures and between creatures extant and extinct. In each comparison, the range of variation is such that only deep time could close the gap. Thus Kelvin's thermodynamic argument against evolution is sidestepped rather than confronted directly. There are, of course, limits to Darwin's fellow feeling with lower beings. In deference to Victorian sensibilities, which are also his sensibilities, Darwin retains from the biblical account the idea that "man" is the pinnacle of creation: just because we might not be God's special creation, he implies, does not mean that we are not special.

At the beginning of *Descent*, Darwin presents us with analogical evidence for the evolutionary forces that link us to our fellow creatures. In each case, analogies demonstrate kinship, while disanalogies point implicitly to deep time, the vast temporal distances required for the evolution of biological complexity. Figure 6.6 illustrates how embryology provides visual evidence for evolution. Even at first glance, in accordance with the Gestalt principle of *Prägnanz*, the dog and human embryos strike us as alike in outline and internal structure. As we examine both more systematically, as we scan and match to compare *d* with *d* or *L* with *L*, eye with eye, or tail with tailbone, this sense of likeness is reinforced. The similarities between the two embryos, similarities that far outweigh any differences, strongly suggest a point of common origin in a distant past; at the same time, the clear evolutionary distance between ourselves and dogs allows readers to infer the vast temporal distance required for this development to take place.

Figure 6.7 depicts a human ear with an anomalous projecting point, a feature that marks, Darwin feels, a reversion to an earlier evolutionary state. We might be inclined to dismiss a character so "trifling" (22) as this projecting point. But, Darwin assures us, this would be an error, as the point is a palpable trace of our distant ancestry, a character that, while unusual

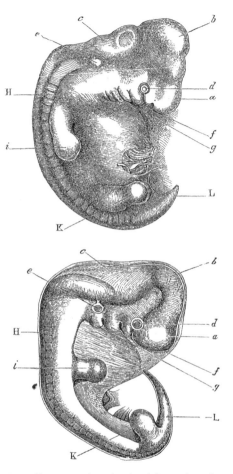

Figure 6.6. Drawings of human embryo (top) and dog embryo (bottom) before the
extremities are apparent (Darwin 1871, 15).

in humans, is universal in our cousins, the monkeys and baboons. The projecting point stands for our evolutionary ancestry; it is a visual index of our
origin. The conclusion is clear. Whether we look at humans in the womb or
at their "reversionary" characters, we see signs of our evolutionary relationship to the lower orders and of our distance from them, a distance whose
temporal dimension must be vast if so much change is to be effected: "by
considering the embryological structure of man—the homologies which he
presents with the lower animals—the rudiments which he retains—and the
reversions to which he is liable, we can partly recall in imagination the former condition of our early progenitors; and can approximately place them

in their proper position in the zoological series" (vol. 2, 389). It is by this means that Darwin traces human ancestry back to an aquatic animal in which both sexes are united in the same individual, a creature half glimpsed in "the dim obscurity of the past" (vol. 2, 389–90), a past the vastness of whose dimensions are deliberately unspecified.

In *Descent*, Darwin employs anthropomorphism as a way of exploring the origin of "mental powers" and "moral sense," signs of our kinship to and distance from other creatures, a distance both evolutionary and temporal. He tells us of a dog that growls and barks fiercely, having noticed the movement of a temporarily abandoned parasol in the slight breeze. "He must," Darwin muses, "have reasoned to himself in a rapid and unconscious manner, that movement with no apparent cause indicated the presence of some strange living agent." From this Darwin infers that "the belief in spiritual agencies would easily pass into the belief in the existence of one or more Gods" (vol. 1, 67). The intellectual leap from a barking dog to the Anglican service is so startling in itself that we might miss an equally startling turn of phrase. Do dogs "reason" to themselves? There is some evidence to support the claim that for Darwin such anthropomorphism is more than a metaphor (Durant 1985). But whatever he might have been willing actually to assert about the human dimension of animals, there is little question that anthropomorphism is a pervasive device that reinforces evolutionary kinship and evolutionary distance at every turn (Darwin 1871, vol. 2, pp. 31, 39, 44, 51, 61, 62, 67, 86, 96, 100, 108–9, 115, 116, 118, 119, 122, 233, 250, 269, 313).

Darwin's *Descent of Man* is mostly verbal; figures 6.6 and 6.7 are its only visuals, though his depiction of the dog and the parasol certainly attests to his powers of evoking pictures by means of words. By contrast, his *Selection in Relation to Sex* relies routinely upon the interaction between

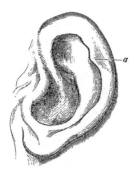

Figure 6.7. Drawing of human ear with anomalous projecting point at *a* (Darwin 1871, 22).

its words and its seventy-four images—the latter mostly comparing and contrasting the physical attributes of males and females within the same species as they pertain to his sexual selection theory. An argument in favor of the interlocking effects of natural and sexual selection relies on the fact that among birds, reptiles, and especially fishes, males "are ornamented with infinitely diversified appendages, and with the most brilliant or conspicuous colours, often arranged in elegant patterns, whilst the females are left unadorned" (vol. 2, 397). For example, Darwin claims that the plumage of the male Argus pheasant has sexual attractiveness as its purpose. For the female Argus pheasant, in other words, the male plumage is symbolic. Again Darwin anthropomorphizes:

> We must conclude that this is the case, as the primary wing-feathers are never displayed, and the ball-and-socket ornaments are not exhibited in full perfection, except when the male assumes the attitude of courtship. The Argus pheasant does not possess brilliant colours, so that his success in courtship appears to have depended on the great size of his plumes, and on the elaboration of the most elegant patterns. Many will declare that it is utterly incredible that a female bird should be able to appreciate fine shading and exquisite patterns. It is undoubtedly a marvelous fact that she should possess this almost human degree of taste, though perhaps she admires the general effect rather than each separate detail. (vol. 2, 92–93)

Darwin subsequently makes his case for the evolutionary descent of these features in the male pheasant over deep time by the careful examination of feathers in living creatures. He begins with a conundrum that attests to his powers of visualization to create pictures with words:

> the ocelli on the wing-feathers of the Argus pheasant, which are shaded in so wonderful a manner as to resemble balls lying within sockets, and which consequently differ from ordinary ocelli. No one, I presume, will attribute the shading, which has excited the admiration of many experienced artists, to chance—to the fortuitous concourse of atoms of colouring matter. That these ornaments should have been formed through the selection of many successive variations, not one of which was originally intended to produce the ball-and-socket effect, seems as incredible, as that one of Raphael's Madonnas should have been formed by the selection of chance daubs of paint made by a long succession of young artists, not one of whom intended at first to draw the human figure. In order to

discover how the ocelli have been developed, we cannot look to a long line of progenitors, nor to various closely-allied forms, for such do not now exist. (vol. 2, 141–42)

Darwin relies for his discovery on a set of visuals (figs. 6.8 to 6.10). While we cannot observe the gradual evolution of the pheasant's ball-and-socket plumage over deep time, Darwin contends, we can reconstruct the process by visually comparing the patterns of the secondary wing feathers in an existing bird. In other words, we can look back—far back—simply by looking closely around us with a trained eye.

As we move from figure 6.8 to 6.10, the ball-and-socket ocelli emerge from the background to the foreground. These Gestalt patterns reveal to us what could plausibly be the pheasant's evolutionary descent. In figure 6.8, we see rows of oblong shapes that converge toward the shaft, "the first trace

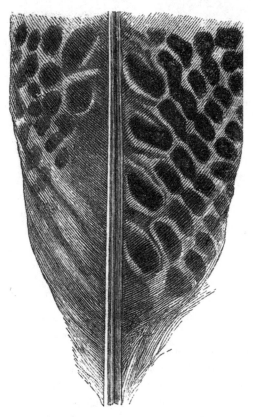

Figure 6.8. Drawing of basal part of the secondary wing feather, nearest to the body of Argus pheasant (Darwin 1871, vol. 2, 144).

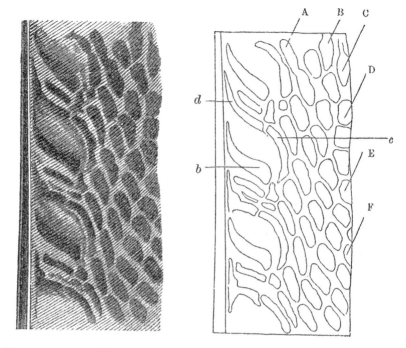

Figure 6.9. Drawing of portion of one of the secondary wing-feathers near to the pheasant body, showing the so-called elliptical ornaments (Darwin 1871, vol. 2, 146).

of an ocellus" (vol. 2, 144) that is "not in any way more remarkable than those of the plumage of many birds" (vol. 2, 145). Figure 6.9 places a realistic representation beside a diagrammatic one, a guide to discerning the components of a middle stage: elliptical ornaments that are halfway between the oblong and the ball-and-socket shapes. The majuscule letters *A* through *F* point to the rows of spots running down to and forming the elliptical ornaments; the miniscule letters *b* through *d* single out the components of one such ornament. Figure 6.10 illustrates the final stage—the completely developed ball-and-socket ornaments. The miniscule letters *a* and *b* point to two perfect ocelli; the majuscule letters *A* through *F* point to dark stripes running obliquely down, each to an ocellus. As we advance from figure 6.8 to 6.10, then, we actually see evolution in progress: from oblong shape to elliptical ornament to ball and socket. These images display what Darwin observed in living creatures; the text provides the training we readers need to decipher their meaning. Darwin concludes this verbal-visual argument with a careful hedge defining the limitations of his claim: "It is obvious that the stages in development exhibited by the feathers on the same bird do not at all necessarily shew us the steps which have been passed through by the extinct progenitors of the species; but they probably give us the clue to the actual steps, and they at least prove to demonstration that a gradation is possible" (vol. 2, 150).

In other words, we need only look closely around us with a trained eye to see what might be remnants of evolution over vast expanses of time. In embryology, in reversionary characters, in the manifestation of mental and moral powers in lower creatures, in comparisons among existing species and between existing and extinct species, we learn from whence we have emerged, and note, presumably with considerable satisfaction, the evolutionary distance we have come.

In *Expression of Emotions in Man and Animals* (1872), Darwin's next

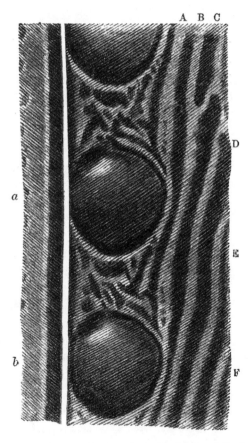

Figure 6.10. Drawing of part of the secondary wing-feather of the Argus pheasant (Darwin 1871, vol. 2, 143).

book, he works out the implications of evolutionary descent in a new arena, emotional expression. It is not altogether clear that animals have emotional states. In *Expression*, Darwin overcomes this difficulty by a strategy already employed in *Sexual Selection*. He anthropomorphizes, consistently endowing birds and animals with emotional states on the basis of their behavior. This imputation enables him to reconstruct the evolution of emotional expression as a gradual progress toward human complexity.

The full set of emotions whose expression Darwin chooses to analyze is problematic. While joy and grief, dejection and despair, love and hatred are indeed emotions, devotion, determination, and patience most assuredly are not. But, as philosopher Amélie Oksenberg Rorty assures us, "emotions do not form a natural class . . . nor can they be sharply distinguished from moods, motives, attitudes, character traits" (1980, 1). While we may not be able to *feel* devotion, determination, or patience, we do express these in various ways. It is their expression only that falls within Darwin's explanatory scope. Darwin's book, then, is united in its concern for the ways in which animals and human beings express moods, motives, attitudes, character traits, and feelings. From these components, we infer states of mind. Darwin's point is consistently that the striking parallels between emotional expressions in human beings and animals strongly suggest evolutionary kinship.

Darwin contends that some emotions and emotional expressions we share with our fellow creatures: joy, affection, anger, astonishment, terror. In the following passage, for example, Darwin visualizes terror, an emotion that reveals itself in the same way in birds, cattle, dogs, cats, monkeys, and human beings:

> With all or almost all animals, even with birds, Terror causes the body to tremble. The skin becomes pale, sweat breaks out, and the hair bristles. The secretions of the alimentary canal and of the kidneys are increased, and they are involuntarily voided, owing to the relaxation of the sphincter muscles, as is known to be the case with man, and as I have seen with cattle, dogs, cats, and monkeys. The breathing is hurried. The heart beats quickly, wildly, and violently; but whether it pumps the blood more efficiently through the body may be doubted, for the surface seems bloodless and the strength of the muscles soon fails. In a frightened horse I have felt through the saddle the beating of the heart so plainly that I could have counted the beats. The mental faculties are much disturbed. Utter prostration soon follows, and even fainting. A terrified canary-bird has been seen not only to tremble and to turn white about the base of

the bill, but to faint; and I once caught a robin in a room, which fainted so completely that for a time I thought it dead. (77; see also 231–32, 241–42, 244, 251, 283)

Supplementing Darwin's vivid depictions of animals in emotional states also experienced by humans—states such as terror, hostility, anger, affection, and humility—are drawings meant to picture the varied visual manifestations of these states. These images, for example, drawings of a terrified cat (128) and man (299), stand for all animals or humans in a particular emotional state. They are at the same time signs of a shared evolutionary path.

A striking parallel between humans and animals also exists in the case of the phenomenon Darwin dubs "antithesis," the tendency for opposite emotions to be expressed by means of antithetical movements. Darwin illustrates this with a contrast: the same dog expressing affection and hostility. It is by means of the Gestalt principles of *Prägnanz* and comparison that we perceive antithesis in figures 6.11 and 6.12, an impressive instance of disjunction. Figure 6.11 represents canine hostility. Figure 6.12, on the other hand, represents friendly feeling. In two contrasting photographs, Darwin illustrates an analogous antithesis in human beings. Figure 6.13 represents indignation; figure 6.14, appeasement. *Prägnanz* and comparison again permit us to see the striking parallels between figures 6.13 and 6.14, and, more important between the two *sets* of figures: 6.11 and 6.12, and 6.13 and 6.14.

Throughout *Expression*, emotional states are imputed to a wide range of creatures, including insects that "express anger, terror, jealousy, and love by their stridulation" (349; see also 56, 77, 85, 88, 91, 94, 114, 123–24, 131, 134, 136, 142, 144, 213). Darwin even goes so far as to attribute consciousness to porcupines:

We can, I think, understand why porcupines have been provided, through the modification of their protective spines, with this special sound-producing instrument. They are nocturnal animals, and if they scented or heard a prowling beast of prey, it would be a great advantage to them in the dark to give warning to their enemy what they were, and that they were furnished with dangerous spines. They would thus escape being attacked. They are, as I may add, so fully conscious of the power of their weapons, that when enraged, they will charge backwards with their spines erected, yet still inclined backwards. (94)

Still, the gap between humans and other creatures is wide. Even when animal and human emotions do not differ, their expression often does.

Figure 6.11. Drawing of dog approaching with hostile intentions (Darwin 1872, 52).

Figure 6.12. Drawing of dog in humble and affectionate state of mind (Darwin 1872, 53).

According to Darwin, dogs express joy by grinning, retracting their upper lip so that the canines are exposed, and the ears are drawn backward. While some speak of this as a smile, Darwin disagrees because he does not see any more pronounced movement of the lips and ears "when dogs utter their bark of joy" (119–20). Humans express this same emotion differently.

Figure 6.13. Photograph of indignant man (Darwin 1872, facing 264).

Figure 6.14. Photograph of apologetic man shrugging his shoulders (Darwin 1872, facing 264).

"When intense, [joy] leads to various purposeless movements—to dancing about, clapping the hands, stamping & c., and to loud laughter. Laughter seems primarily to be the expression of mere joy or happiness" in humans (196). Finally, there are emotions that only humans experience, a firm indication that we have transcended our evolutionary origins. Blushing is the paradigmatic expression of a cluster of emotions only humans feel because only they are not only conscious but also self-conscious:

> blushing—whether due to shyness—to shame for a real crime—to shame from a breach of the laws of etiquette—to modesty from humility—to modesty from an indelicacy—depends in all cases on the same principle; this principle being a sensitive regard for the opinion, more particularly the depreciation of others, primarily in relation to our personal appearance, *especially of our faces; and secondarily, through the force of association and habit, in relation to* the opinion of others on our conduct. (335–36, our emphasis)

The major theme of *Descent and Selection* (1871) and of *Expression* (1872) is the same: "species are modified descendants of other species" (1871, 2). With the exception of *Descent,* the first portion of the first book, the arguments Darwin makes are consistently the product of verbal-visual interaction. Both books create a world in which deep time must be presupposed if we are to recognize our kinship with our fellow creatures and our distance from them. In *Descent and Selection,* our many similarities with other creatures permit the inference of evolutionary origin; in *Expression,* a consistent anthropomorphism permits Darwin to reconstruct the evolution of emotional expression as a conclusive demonstration of our affinity with our fellow creatures, the living representatives of our faunal ancestry. At the same time, Darwin is careful to emphasize the disanalogies between our emotions and emotional expressions and those of our fellow creatures. Employed thus, his anthropomorphism serves an overall strategic purpose: to domesticate deep time by shortening the distance between us and our fellow creatures, while at the same time taking care to flatter us for having reached the pinnacle of creation, most evident in our mental and moral powers and best exemplified in blushing, a clear index of those powers.

ANTHROPOLOGICAL DEEP TIME

In *Pre-Historic Times,* John Lubbock (1875) dismisses the limit set by Bishop Ussher on human existence, a temporal space barely extensive enough to accommodate human history. As far as he is concerned, "our belief in the antiquity of man rests not on isolated calculations, but on the changes which have taken place since his appearance; changes in geography, in the fauna, and in the climate of Europe. Valleys have been deepened, widened, and partially filled up again; caves through which subterranean rivers once ran are now left dry; even the configuration of land has been materially altered, and Africa finally separated from Europe" (419). It is these changes that permit us to have an accurate view of the vastness of our long past, a deep time that, if properly informed, we can intuit merely by scanning the valley of the Somme from the heights of neighboring villages:

> What date are we to ascribe to the men who lived when the Somme was beginning its great task [of excavating its valley]? No one can properly appreciate the lapse of time indicated, who has not stood on the heights of Liercourt, Picquigny, or on one of the other points overlooking the valley: nor, I am sure, could any geologist return from such a visit without an overpowering sense of the changes which have taken place, and the length of time which must have elapsed since the first appearance of man in Western Europe. (384)

While Lubbock cannot securely date the artifacts with which he makes his case for human antiquity, he can order them according to successive stages of prehistory: the Old Stone Age, characterized by rough-hewn implements, the New Stone Age, characterized by their well-hewn counterparts, and the Bronze Age, which reaches to, and in some cases overlaps, the edge of recorded history. There are other artifacts that contribute to our understanding of our prehistoric ancestors. These protohumans used clay to make pottery; they engaged in agriculture and domesticated animals; they housed and clothed themselves; they buried their dead. Their artists painted and etched in stone or bone; their craftsmen forged and formed ornaments. Moreover, from their unearthed skulls, their relative intelligence can be inferred; from fossilized flora and fauna, their climate. Together, these form a cluster by means of which their way of life and degree of civilization—what we would call their culture—may be derived.

Fully to domesticate deep time, however, to accommodate to it emotionally, Victorians needed more than arguments based on artifactual and fossil evidence; they needed to understand their remote ancestors as kin as well as kind. To dramatize this kinship, to make it real for his audience, Lubbock employed images that permitted his readers to become virtual witnesses to prehistory. There are 161 of these, or nearly one for every three

pages. These are not imaginative reconstructions of the distant past as in a
museum diorama, but realistically rendered pictures that establish analogies between prehistoric ways of life and those of "primitive" people who
are the Victorians' contemporaries:

> If we wish clearly to understand the antiquities of Europe, we must com-
> pare them with the rude implements and weapons still, or until lately,
> used by the savage races in other parts of the world. In fact, the Van
> Diemener and the South American are to the antiquary what the opos-
> sum and the sloth are to the geologist. (428)

Lubbock is aware of the limits of analogy. He knows that Stone Age civilization differs not only in degree but in kind from that of surviving "savage"
races. Nevertheless, he finds the evidence conceptually cogent because it is
so visually striking. One instance of the compelling character of this evidence arises out of a visual comparison between a contemporary Fuegian
harpoon and its prehistoric counterpart (fig. 6.15).

These two harpoons are separated by eons of human history. When
viewed together, however, time, in effect, vanishes. Gestalt comparison
permits us to perceive interchangeability and to infer from that a common
purpose: in each case, the downward-sloping blades are formed to stick fast
in a wounded prey. We can *see* the Fuegian artifact as an index of a Fuegian
practice; we can also *look past* this artifact in order to perceive a habitual
practice of our prehistoric forbears. Throughout *Pre-Historic Times*, Lubbock domesticates time by means of such analogies: analogies of tools and
weapons (84, 95–96, 507, 477); analogies of canoe building and food preparation (185, 327); analogies of ornaments and art (37, 44, 335). He even goes
so far as to analogize ways of life. Of the prehistoric cave dwellers of the
Dordogne Valley, he says, "so far as the present evidence is concerned, it
appears to indicate a race of men living almost as some of the Esquimaux
now do, and as the Laplanders did a few hundred years ago" (336–37; see
also 241).

From our observations of the everyday life of primitive peoples, we can
also infer by analogy the means by which artifacts were manufactured by
prehistoric humans. Figure 6.16 displays a prehistoric flint core and flint
flakes chipped from such a core. From experimentation, we know that
if a blow is struck at an angle against the flint core, the fracture eventually forms a bladelike flake with a small bulb or projection. Such a flake
is marked *a* in the image that is second from the right. But we do not have

Figure 6.15. Drawings of contemporary Fuegian harpoon (left side, Lubbock 1875, 540) and prehistoric harpoon (right side, Lubbock 1875, 105).

to experiment: we can see for ourselves. Figure 6.17 depicts contemporary Australian natives making such flakes.

It is crucial to Lubbock's evolutionary thesis that all contemporary "savage" races are not at the same level of progress toward civilization, that they differ significantly on such measures as weaponry and fortifications, fishing, agriculture, and domestication. In the table reproduced in figure 6.18, as we shift our attention from Bushmen and the natives of Tierra del Fuego to those of Feegee and Tahiti, we see a variety of primitivenesses; we also see, under Lubbock's guidance, stages in earlier human progress. In effect, we discern in these comparisons the missing links between our apelike remote ancestors and ourselves. Measured in these terms, levels of civilization vary widely from race to race. Bushmen have weak bows and arrows that are useless in warfare and use dogs for hunting. They have no pottery

and no agriculture. They have no redeeming features whatever: "Bleek regards them as the lowest of the human races and Haeckel even goes so far as
to assert that they seem 'to the unprejudiced comparative student of nature,
to manifest a closer connection with the gorilla and the chimpanzee than
with a Kant or a Goëthe'" (434–35). Nonetheless, whatever their degree of
civilization, it is Lubbock's view that none of these peoples is wholly to be
admired. Feegians have good bows and arrows, slings, pottery, very good
canoes, agriculture, fortifications, fishhooks and nets, and they domesticate
dogs for food. Of the Feegians, Lubbock quotes the view of Captain John El-
phinstone Erskine with approval, a view in which admiration is tempered
with distaste: "On contemplating the character of this extraordinary people,

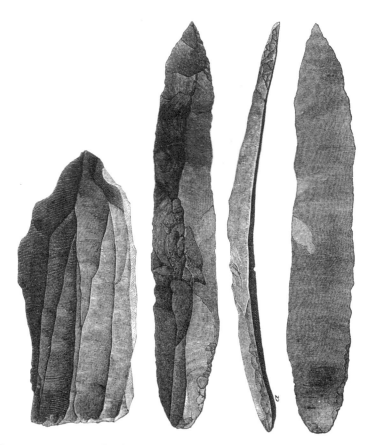

Figure 6.16. Drawings of prehistoric artifacts (Lubbock 1875, 85). From left to right, a
flint core or nucleus and three flint flakes of Danish origin.

Figure 6.17. Drawing of Australian natives making flint flakes (Lubbock 1875, 88).

the mind is struck with wonder and awe at the mixture of a complicated and carefully-conducted political system, highly finished manners, and ceremonious politeness, with a ferocity and practice of savage vices which is probably unparalleled in any other part of the world" (460).

In effect, contemporary primitives form a living museum of human evolution. Thanks to them, anthropological time and our time can be made roughly to coincide, a feat of reconstruction Lubbock vividly depicts:

Much as still remains to be made out respecting the men of the Stone period, the facts already ascertained, like a few strokes by a clever draughtsman, *supply* us with the elements of an outline sketch. Carrying our imagination back into the past, we *see* before us on the low shores of the Danish Archipelago a race of small men, with heavy overhanging brows, round heads, and faces probably like those of the present Laplanders. As they must evidently have had some protection from the weather, it is most probable that they lived in tents made of skins. The

	Easter Islanders	Fuegians	Bushmen	Hottentots	Andamaners	Australians		Esquimaux		North American Indians		New Zealanders	Fergelans	Society Islanders	Friendly Islanders
						North-East	West	Southern	Northern	West	East				
Bows and Arrows	...	Weak	Weak	Weak	Good	...	...	Good	...	Good	Good	...	Good	Weak	Weak
Slings	...	Yes	...	...	...	...	...	Yes	...	Yes	Yes	...	Yes	Yes	?
Throwing-Sticks	...	...	...	...	...	Yes	Yes	Yes	?	...	...	...	...	...	...
Boomerangs	...	...	...	...	...	Yes	Yes	...	...	...	...	...	...	...	...
Bolas	...	...	...	...	...	...	...	Yes	?	...	...	...	...	...	...
Pottery	...	...	...	Yes	...	...	...	...	...	...	Yes	...	Yes	...	...
Canoes	Bad	Bad	...	...	Good	...	Bad	Good	...	Bad	Middling	Very good	Very good	Very good	Very good
Agriculture	...	...	...	...	...	...	...	...	...	...	Maize	Yes	Yes	Yes	Yes
Fortifications	...	...	...	...	...	...	...	...	...	...	...	Many	Yes	...	Yes
Fish-hooks	...	Stone	...	Iron	?	...	Neat	Bone	...	Yes	Yes	Bone and shell	Bone and shell	Bone and shell	Shell
Nets	...	...	...	Yes	Good	...	Neat .	Small	For bird catching	Yes	Yes	Large	Yes	Large	Yes
Dogs	...	For hunting.	For hunting.	For hunting.	...	For hunting.	For hunting.	For hunting.	For draught.	For wool & hunting.	For hunting.	For food	For food	For food	...
Hogs (Domestic)	...	...	...	...	...	...	...	...	...	...	...	...	Some	Many	...

Figure 6.18. Tabular comparison of the measure of civilization of the various savage tribes (Lubbock 1875, 553).

total absence of metal in the Kjökkermöddings [mounds] indicates that they had not yet any weapons except those made of wood, stone, horn, and bone. (240–41; see also 379 and 383–84; emphasis ours)

Both of the predicates we have italicized, "supply" and "see," are in the present tense. But "supply" refers to perception, while "see" refers to the visual imagination.

It is this coincidence of perception and visual imagination that licenses an anthropological narrative that Lubbock has built on well-founded conjectures inferred from fossils, prehistoric artifacts, and contemporary primitive people. It is a narrative driven by "the great principle of Natural Selection, which is to biology what the law of gravitation is for astronomy" (593). In Lubbock's eyes, however, natural selection has been transformed from the force it once was, a force indifferent to human purposes. It is now an entirely positive force that permits us not only to reconstruct the past of humankind, but to predict for it a future filled with hope. This is possible— indeed inevitable—because natural selection has finally permitted humankind to escape its own worst consequence, the continued struggle for existence. External conditions are now largely in our control, a result that has in recent times culminated in "the wonderful intellect of the Germanic races" (591; quoting Alfred Russel Wallace). Far from being too good to be true, Utopia "turns out to be a consequence of natural laws, and once more we find that the simple truth exceeds the most brilliant flights of the imagination" (603; see Stocking 1982, 40–41, 77–78, 88–89, 131–32). It is thus that an argument heavily dependent on verbal-visual interaction achieves an apotheosis perfectly in tune with Victorian sensibilities.

CONCLUSION

Geikie, Darwin, and Lubbock are all in the same line of work: through words and images they create an argument and a narrative in which deep time is rendered acceptable to a broad Victorian audience, scientists and nonscientists alike. By means of verbal-visual interactions, they repeatedly transform images of present-day landscapes, of natural and man-made objects, and of living beings into signifiers of a much different past shaped by natural laws operating over immense stretches of time.

Nevertheless, their efforts vary in persuasiveness. This difference can be most perspicuously conceived as a deviation into the rhetorical, in its pejorative sense. In this respect, Geikie comes out best. Although he casts his book in the form of a mystery, this attempt at audience accommodation does not touch the core of his work: the signs in the landscape he points to genuinely are effects, and the glaciers he imagines genuinely are causes. Properly interpreted, the images he employs actually represent both the present and the deep past. Lubbock is on less-secure ground. True, he can justify his persistent analogical resort by reference to genuine visual similarities in tools and weapons between contemporary indigenous peoples and prehistoric ancestors; true, he is aware and makes us aware of the disanalogies that exist between these two groups; true, in natural selection he has a plausible explanation for differential development. But while he recognizes that differences among tribes are due to such external conditions as climate, vegetation, and food sources (550–51), he infers impermissibly that nevertheless the indigenous peoples he examines are uniformly victims of arrested development. But one cannot infer on evolutionary grounds that any tribe—including the Victorian tribe—ranks above another. Each is presumptively well adapted to its environment.

In the matter of plausibility, Darwin fares worst. While the evolutionary arguments in *Descent* hold up, his consistent anthropomorphism in *Expression* is hard to take seriously. *Expression* is also undermined because some of its photographs are clearly posed, and there is evidence that the some woodcuts were "enhanced" at Darwin's request. Neither can count as scientific evidence in our sense or, perhaps, in his (Browne 1985, 313–14; Browne

2002, 362–69). There is another problem both Darwin and Lubbock share: it is no accident that the overall story they have to tell is a tale in which white upper-middle-class Englishmen are, apparently, by natural right the pinnacle of God's creation. Sight can lead to insight to be sure, but it is never free of cultural bias.

So far in this book we have dealt with images limited by the constraints of the printed page. In the two chapters that follow, the final two chapters of the book, we concern ourselves with the ways in which verbal-visual inter-action evolved in adapting to two new media: PowerPoint and the Internet. We will argue that this migration is engendering a revolution, both in the communication of science and in its representation.

CHAPTER SEVEN

The Public Science Lecture:
PowerPoint Transforms a Genre

The lead Genentech researcher took the audience [American Society of Clinical Oncology] through one slide after another—*click, click, click*—laying out the design and scope of the study, until he came to the crucial moment: the Kaplan-Meier. [These are graphs that dramatize the difference between drugs that work and those that do not by comparison of survival curves for experimental and control groups.] At that point, what he said became irrelevant. The members of the audience saw daylight between the two lines, for a patient population in which that almost never happened, and they leaped to their feet and gave him an ovation.

—Malcolm Gladwell, "The Treatment," *New Yorker* (May 17, 2010, 69)

POWERPOINT'S USES AND ABUSES

On December 7, 2007, at the Karolinska Institutet in Stockholm, Martin Evans, Oliver Smithies, and Mario Capecchi—newly minted laureates in Physiology or Medicine—delivered their Nobel lectures, each running about forty minutes.[1] All three presented slide shows explaining their roles in the development of key technologies for the manipulation of genetic sequences.[2] All are practiced speakers and gave polished and engaging talks.

---

1. One of the problems in analyzing a computer file of a scientific PowerPoint presentation is that one is not usually able to hear the spoken words that explain each slide and thus must guess what the speaker might have said. Thanks to the Nobel website (http://nobelprize.org /nobel_prizes/medicine/laureates/2007), however, we were able to gain access not only to the speakers' slides, but also to videos and transcripts of their talks.

2. There are other presentation programs besides PowerPoint on the market, and we do not know for sure which one was used for the examples in this chapter. We use the term "Power-Point" as a catch-all term for any such program.

But only one of the three took full advantage of the communicative possibilities afforded by PowerPoint, namely, its ability to place the visual steadfastly at the center of scientific communication. In Evans's and Smithies's lectures—traditional talks supplemented by slides—audiences were confronted with two competing verbal streams: the steady flow of speech coupled with the relentless march of slides stuffed, even overstuffed, with words. Evans's slides are filled with bulleted lists: over half contain such lists. In addition, some of his slides reproduce pages from key articles; others, pages from his laboratory notebooks. Smithies has a slightly different approach: for the most part, his slides (over ninety of them, or approximately one for every half minute of speaking time) reproduce the lab notebook entries made on the days of his important discoveries. Figure 7.1 is an example. The words and numbers these entries contain are barely legible; in any case, their significance would be difficult to discern without detailed explanation that would

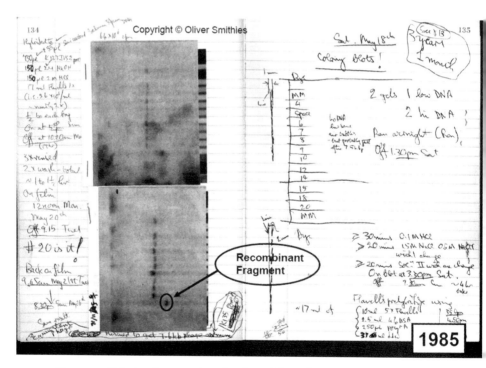

Figure 7.1. Slide showing pages from Smithies's lab notebook, May 18, 1985 (Smithies 2007). The resulting journal article on gene targeting was central to award of Nobel Prize. Onto the image Smithies superimposed notebook year and highlighted the visual blot that departed from those above it, indicating success (subclone "#20"). © The Nobel Foundation.

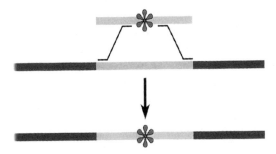

Figure 7.2. Slide showing regeneration of a functional neomycin-resistant gene (neo^{r+}) by gene targeting (Capecchi 2007). Homologous recombination of the two defective neo^{r-} genes (upper two lines) yields a functional neo^{r+} gene (bottom line). This technique is used by molecular biologists in establishing mammalian cell lines. © The Nobel Foundation.

only distract from the main points in his lecture. While each slide gives us a glimpse into the messy process of scientific discovery, none is really helpful in understanding the science involved, at least within the course of a forty-minute lecture.

Capecchi's communicative approach is different from that of his fellow laureates. He presents his discoveries on gene targeting in a series of thirty-five PowerPoint slides that maximize the visual and minimize supporting text; in most cases, as in figure 7.2, his slides lack even descriptive headlines. In all cases, words, parsimoniously employed, keep both speaker and audience on track. In Capecchi's hands, we see the transformation of a genre, the public science lecture. Words govern the intellectual shape of his talk, while visuals—photographs, graphs, and diagrams—form its intellectual substance. Meaning is the product of verbal-visual interaction.

Capecchi builds into his presentation an understanding of the differences between interlocutors, readers, and audiences. Interlocutors not only listen, they continually question and respond. Audiences do not have this option. Moreover, audiences lack the flexibility open to readers—they cannot vary the pace; they cannot re-listen. It is lecturers who must compensate for these deficiencies, accommodating their talks to the well-known constraints that verbal complexity and working memory place on audience comprehension. How do lecturers approach this challenge? Some write their lectures out and read them aloud. But if they do, they are misusing a communicative technology designed specifically to accommodate silent reading. Moreover, if they read their lectures aloud, unless they have committed them to memory, their attention will be divided between the audience and the page. This is not what audiences expect from lecturers; they want those who talk to them to maintain eye contact, sustaining the illusion that each is the lecturer's sole addressee. To avoid the problem of divided attention, lecturers can choose simply to talk to their audiences. But if they do, they risk losing their way; moreover, they open the door to hesitation, verbal tics, and prolixity. Properly employed, PowerPoint can help solve these problems, keeping both lecturers and audiences on target (Kosslyn 2007). It can do so by limiting the verbal content of slides, centering them whenever possible on the visual.

Because this centering is routinely ignored, PowerPoint has been accompanied by extensive criticism, most prominently the tidal wave generated by Edward Tufte. In *The Cognitive Style of PowerPoint* (2003a), Tufte makes a strong case that PowerPoint is a widely used software program whose bullet-point formats, when blindly accepted, lead to presentations that confuse and mislead rather than enlighten. He holds PowerPoint responsible not only for the *Columbia* disaster but also far more widespread communicative disasters in business meetings and lecture halls: "At a minimum, a presentation format should do no harm. Yet the PowerPoint style routinely disrupts, dominates, and trivializes content. Thus PowerPoint presentations too often resemble a school play—very loud, very slow, and very simple" (Tufte 2003b).

Tufte's disdain of PowerPoint notwithstanding, bullet-point slides are now omnipresent in the lecture hall. On the basis of a quantitative analysis of more than one hundred PowerPoint presentations, Garner et al. (2009) concluded that the slides were dominated by three basic formats, all typically having noun phrase (sentence fragment) headlines at the top. First and most common was a headline with a bulleted list or lists below (39%). Second was a headline, underneath of which appeared an image or images and a bulleted list, side by side (28%). Third was a headline with an image or images below it and no bullets (23%).[3] Thus almost 70% of the slides had a bulleted list.[4] Tufte is rightly contemptuous of the ubiquity of such lists: "Medieval in its preoccupations with hierarchical distinctions, the PowerPoint format signals every bullet's status in 4 or 5 different simultaneous ways: by the order in sequence, extent of indent, size of bullet, and size of type associated with various bullets" (2004a, 12).

Alley and Neeley (2005), Alley et al. (2006), Neeley et al. (2009), and Markel (2009) have been strong advocates of an alternative: presentations

3. Other structures compose the remaining 12% of slides.
4. We assume here Garner el al. (2009) lumped together numbered with bulleted lists. For the purposes of our later discussion, we consider them essentially equivalent except that numbers are used to indicate order of priority.

that favor a slide format in which a headline captures the main point in a complete sentence, underneath which is a supporting image accompanied by minimal text. This format dispenses with the bulleted list entirely. For oral presentations, such as the ones we examine in this chapter, this structure generally works well. It is Capecchi's way. To apply it without exception, however, would be draconian. Sometimes—in introductory or concluding sections, for example—bullet points are the best solution. Moreover, the Alley and Neely format would not be appropriate for PowerPoint lectures designed to be stored on the web after delivery or emailed to individuals unable to attend the lecture (Farkas 2006). Such presentations would be difficult to decipher absent more verbal scaffolding than oral presentations require. The answer might be two versions of the same presentation, one for oral presentation, one for silent reading—though it seems unrealistic to expect most scientists to prepare these. We maintain, however, that a slide with bullet points can be suitable for both purposes. We think this is the case with figure 7.3, a slide from Thomas Seeley's presentation, a lecture to be analyzed in detail later in this chapter.[5]

In the remainder of the chapter we will discuss four PowerPoint lectures, chosen because each has exploited PowerPoint's advantage: the centering of the visual. The first two presentations are aimed at audiences other than strictly professional. The first, "The Ongoing, Mind-Blowing Eruption of Mount St. Helens," is by Daniel Dzurisin of the Cascades Volcano Observatory, United States Geological Survey. It is an example of a PowerPoint talk for "science enthusiasts," that is, anyone with a strong interest in volcanoes. No highly specialized expertise in volcanology is presupposed. The second is "House Hunting by Honey Bees: A Study in Group Decision-Making." It is by Thomas D. Seeley of the Department of Neurobiology and Behavior, Cornell University, who took this slideshow on the road before an audience of undergraduate science students and their teachers.

The two lectures we analyze next are meant for strictly professional audiences. The first is Richard Iverson's "A Dynamic Model of Seismogenic Volcanic Extrusion, Mount St. Helens, 2004–2005." Like Dzurisin, Iverson works for the Cascades Volcano Observatory. In arguing for the validity of a newly created mathematical model, he presupposes an audience of experts thoroughly familiar with the science and graphic conventions in play and easily able to process complex visual information with rapidity. Our final example is "Integrating Ecological and Evolutionary Studies of Biological

---

5. And frankly, since such bulleted slides also help keep the speaker on point, we do not see this slide style becoming extinct any time soon.

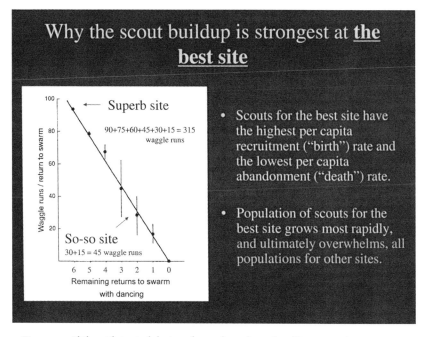

Figure 7.3. Slide with typical design of complete-phrase headline at top, image on one side, and bulleted list on the other (Seeley 2005). Illustrates the point that scout bees in favor of a "superb site" for a new nest are far better at recruiting their fellow bees to their way of thinking.

Diversity: Patterns and Processes of Adaptive Radiation in Caribbean Lizards," by Jonathan Losos of Harvard University's Department of Organismic and Evolutionary Biology. Losos is concerned with the profound effects of deep time on a single species; he is also concerned with more rapid adaptation in response to a changed hostile environment. Losos's lecture differs from that of Iverson's in one important respect: while Iverson focuses on the product, Losos also focuses on the process of scientific investigation. In so doing, he reverses a long-term trend in communicating science. All four lectures are alike in that their meaning is a product of verbal-visual interaction.

DZURISIN'S PRESENTATION FOR A
GENERAL AUDIENCE: NARRATIVE

To structure his general-audience lecture, "The Ongoing, Mind-Blowing Eruption of Mount St. Helens," Daniel Dzurisin relies on narrative, that is, a sequence of events exhibiting a "grammar" that consists of "setting, theme, characters, plot, and resolution" (Dymock 2007, 162; see also Price, Roberts, and Jackson 2006). On reflection, the resort to narrative is unsurprising. What Cliff Atkinson intuits in his impressive *Beyond Bullet Points* (2008), researchers in developmental psychology had already discovered through painstaking experimental work: storytelling and story understanding are basic human capacities on which speakers can consistently rely. These develop soon after language emerges: children as young as four have a fair grasp of the elements of story grammars (Dymock 2007, 184) and adults never leave storytelling far behind. Narrative is also a way of making sense of lives, our own and others'. According to Jürgen Habermas (1987), narrative is the primary vehicle for both self-understanding and social integration:

> Narrative practice not only serves trivial needs for mutual understanding among members trying to coordinate their common tasks; it also has a function in the self-understanding of persons. They have to objectivate their belonging to the lifeworld to which, in their actual roles as participants in communication, they do belong. For they can develop personal identities only if they recognize that the sequences of their own actions form narratively presentable life histories; they can develop social identities only if they recognize that they maintain their membership in social groups by way of participating in interactions, and thus that they are caught up in the narratively presentable histories of collectivities. Collectivities maintain their identities only to the extent that the ideas members have of their lifeworld overlap sufficiently and condense into unproblematic background convictions. (136)

Habermas's son, Tilmann, a developmental psychologist, gives his father's philosophical insight psychological substance. Based on experimental studies, he asserts that "even pre-school children show some competence in the relatively easy task of assigning events to life phases, which is mastered by age 13. . . . The more difficult task of sequencing life-events is mastered by the end of adolescence" (Habermas 2007, 4).

Dzurisin's public lecture on the Mount St. Helens eruption fulfills the expectations of the standard story grammar. It has as its setting the state of Washington, where the volcano is located. Its "main character" is the team of volcanologists in charge of monitoring the eruption; their response to the impending event is to intensify their efforts and to keep the general public informed of the eruption's progress. That is the plot of Dzurisin's story. The resolution of this plot occurs when the eruption is over, the public having been kept informed and safe. That public funds have been well spent in the

Figure 7.4. Slide picturing lava dome that resulted from the 2004–2006 Mount St. Helens eruption.

public's interest is the story's theme. In a sense, Dzurisin's story has been told only so that this theme may be inferred. This thematic use of narrative is especially evident in popular persuasion: in Aesop's fables, in Christ's parables, in the exempla of sermons, and in such children's tales as *The Little Engine That Could.*

The slides of the first half of Dzurisin's presentation permit the audience to participate vicariously in the event, to sit in a front-row seat beside the volcanologists as they monitor the eruption. In the second half of the presentation, the emphasis shifts: the volcanologists and their equipment are now center stage; as these scientists monitor the eruption, the audience looks over their shoulders. Only in this way can the audience appreciate the complexity and difficulty of the scientists' task.

The story of the eruption is told in flashback: to dramatize its impact on the audience, Dzurisin begins with the impressive result, figure 7.4, the new lava dome rising out of the volcano crater like a whale from the ocean. How was this dome created? The rest of his presentation answers that question. In order to show the ways in which the verbal and the visual interact to convey a single narrative message, we analyze the whole of Dzurisin's initial episode, a sequence that begins with figure 7.4, the photograph of the new lava dome. In contrast to the PowerPoint format recommended by Alley and Neeley, this slide has two "headlines," one at the top, the other at the bottom. This deviation, however, has a clear communicative purpose, beginning in medias res. The top headline is designed to focus the audience's attention on the arresting photograph. From there the audience's gaze descends past that photograph ("2004–2006 lava dome") to a comment showing empathy with their probable reaction. By means of verbal-visual interaction, the image has become an integral part of the grammar of the story.

The following slide, figure 7.5, takes us back to the story's origin, September 2004. The slide design consists of two graphs placed side by side, with a complete sentence headline above them, explaining their significance. In it, the audience sees what the volcanologists saw at their moment of discovery, a break in routine as the volcano's "heart-beat" suddenly quickens. At first, the audience sees only the left-side graph; next it sees both sides, experiencing vicariously the dramatic contrast the volcanologists experienced when monitoring the surface electronic properties (SEP) over two days. The slide's verbal accompaniment mimics the dramatic change the two graphs depict: as the graphs reveal their secret, the sentence reveals its full significance.

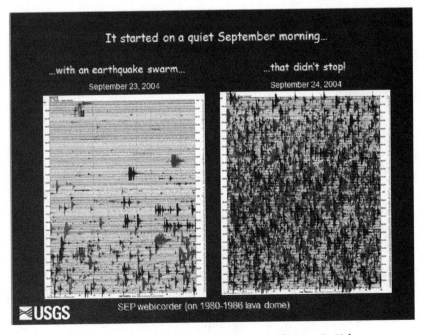

Figure 7.5. Slide indicating onset of the eruption of Mount St. Helens.

Dzurisin begins with an independent clause headed by a free-floating initial pronoun, "it," followed by a phrase that gives us the first clue to *its* identity. A dependent clause follows, cast, significantly, in the emphatic negative (compare " . . . that didn't stop" with the bland " . . . that continued"). In a bit of additional mimicry, each grammatical unit is linked by ellipses, a visual realization of the volcanologists' reaction as they hold their collective breath in the face of an exciting and disturbing anomaly. Finally, the sentence ends in an exclamation point, a mark of punctuation that heads the graph on the right, the solution of the mystery of the initial referent: figure 7.5 is linked to figure 7.4 as cause to effect. In the language of enhanced dual coding theory, figure 7.5 is indexical of seismic activity, while figure 7.4 is iconic, representative of a massive lava dome ("as long as the Eiffel tower") that sometimes forms in a volcanic eruption. It is unlike anything known to have occurred previously on Mount St. Helens.

In figure 7.6, the next slide in the sequence, the shift over seven days (09/24 to 09/30) from mild to intense seismic activity is mimicked in a sentence that shifts typeface from roman to italic. In addition, the use of the past progressive tense—saying that several earthquakes "were occurring"— indicates that the action may extend into a future of continuing and

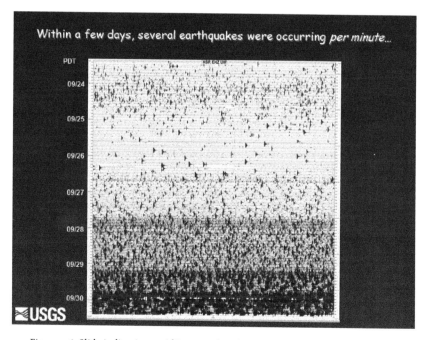

Figure 7.6. Slide indicating rapid increase in seismic activity over a period of days.

Figure 7.7. Slide picturing welt forming as a result of the earthquakes.

intensifying earthquakes. The visual gives scientific meaning to the verbal: the graph is a realization of the words Dzurisin italicizes: *"per minute."* In addition, there are verbal links from slide to slide. The phrase "within a few days" links this slide to the previous one, as does the fact that Dzurisin uses the same format to illustrate the rapid buildup of seismic forces. Moreover, this slide, like its partner in figure 7.5, is linked to the first slide in the sequence as cause to effect.

Figure 7.7, the next slide in the sequence, shifts from one visual form to another: from the graphic evidence of increasing seismic activity to photographic evidence that shows the evolving effect of these "drum-beat" earthquakes on the shape of the mountain. In this slide, the date in the right-hand corner, "September 29, 2004," links to the previous slide's "09/29," the date that the frequency of these earthquakes began rapidly to increase. At the same time, the ellipsis in this clause links it to the sentence on the previous slide, forming a complete sentence. The tense in both of these sentences is the same—the past progressive, a choice that is particularly apt in the case of figure 7.7 because its image must be seen as a "freeze-frame," one instance in a continuous process. These last three slides (figs. 7.5 to 7.7) form a single coherent unit, a parallel sequence of cause and effect: first, the

October 1, 2004

Mount St. Helens' first eruption of the 21st century was underway!

USGS

Figure 7.8. Slide picturing eruption in progress.

drum-beat earthquakes are detected and intensify; then the welt appears, symbolizing an impending volcanic explosion.

Figure 7.8, the next slide, completes the sentence begun at the bottom of the previous slide; at the same time, its image is linked to the previous one as cause to effect: a large welt forms on September 29, followed by an eruption on October 1. This slide ends the first narrative sequence: its billowing smoke and ash unmistakably reveal the dynamic process the graphs have detected and displayed. The process foretold in the progressive tenses of the previous slides now unfolds before our eyes. In this initial sequence, the parallel between the verbal and the visual has been consistently realized in twin semiotic chains. The words in this final slide in the sequence complete the sentence begun on the previous slide; at the same time, they sum up the whole of the first episode. The photograph provides the visual climax of this episode. Audiences now understand the eruption as a calculable nexus of forces, realized through verbal-visual interaction.

Figure 7.9, the final slide of Dzurisin's with which we deal, marks the transition from the first narrative sequence to the second, from the story of the eruption to the story of keeping the public informed of its progress. In a full sentence, its title announces the new topic: the communication of the seismic results gathered for the media and the public by the United States Geological Survey (USGS) and the Pacific Northwest Seismic Network (PNSN). The accompanying graph is linked to the previous slide sequence by the box on the left labeled "Notice of volcanic unrest." The labels in the inset graph on the right link science in the making to making science public.

The graph in the slide is indexical: the jagged graph line, colored red in the slide, traces seismic activity over a long period, and the vertically placed words specify the effect it caused. The full graph may seem too overcrowded for general public understanding. The audience, however, sees the slide unfold in several stages. First, they see only the slide title and the long trace of seismic activity spanning the time from September 22, 2004, through January 24, 2005. The boxes on the left and right are not yet visible. This first stage clearly shows major volcanic activity, followed by a gradual quieting. It presents the big picture. In the second stage, a box appears on the left of the screen with the words "Notice of volcanic unrest (9/26)." Here the lecturer focuses the audience's attention on the major volcanic activity from September 29 through October 14. In the next stage, the focus turns to the

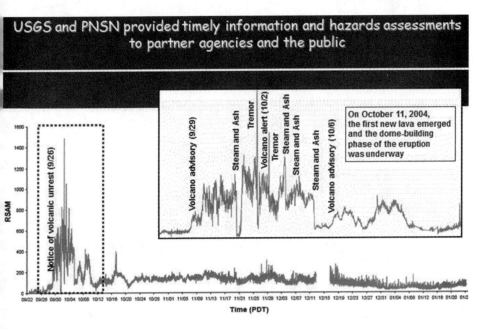

Figure 7.9. Slide showing the onset of the eruption of Mount St. Helens. The graph plots real-time seismic amplitude (RSAM) as a function of time in Pacific Daylight Time (PDT).

box on the upper right of the screen, a version of the box on the left with labels added that interpret the meaning of the changing curve. The audience sees the text for each date in three stages (9/29, 10/2, 10/6), allowing the lecturer to focus on one date at a time. In effect, this sequencing helps the viewer perform the necessary scanning and matching to make sense of the graph. As a consequence of this sequencing, the audience sees not a cluttered graphic, but a sequential signal, an unfolding story: thanks to the precise monitoring of seismic activity, USGS and PNSN were able to provide timely information on the eruption to the media and public.

Dzurisin's story continues but we will stop here. Dzurisin presents a clear and compelling story that integrates the verbal and the visual into a single message adapted to his general audience. Because his primary goal is that his audience reexperience what he experienced, narrative is his medium, a story told in parallel words and images—alternating between pictures of the natural world in turmoil and the data trends underlying that turmoil. Throughout, the crucial unit of analysis has not been the individual slide—the focus of most past pedagogical literature on PowerPoint—but the extent to which the individual slide is integrated into the lecture as a whole.

SEELEY'S PRESENTATION FOR COLLEGE AUDIENCE: NARRATIVE AND ARGUMENT COMBINED

When we move from Daniel Dzurisin's general-audience lecture on the Mount St. Helens eruption to Thomas Seeley's on decision making among honeybees, we shift from telling a story about nature to addressing a series of research problems—the more typical job of scientific lectures. The technical expectations of the audience are somewhat higher here; we have shifted from a diverse group of science enthusiasts to an audience of college teachers and their students with a prior interest in apian behavior. The practical problem for the bees is this. A swarm has grown too numerous for their hive and must choose a new site to colonize. Scout bees, making up about five percent of the hive's population, have the job of finding the new home. How is this important decision made? Four related research problems are addressed in this presentation:

- Do the scouting bees reach a decision by consensus or by quorum?
- How is the decision of the scouts conveyed to the rest of the swarm?
- Is that decision the best choice?
- Can the bees teach us anything about our own decision-making?

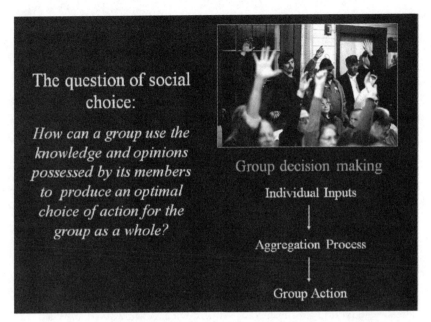

Figure 7.10. The opening content slide of "House Hunting by Honey Bees."

Let us turn first to figure 7.10, the slide presented just after Seeley's title slide. In this second slide, he poses the central question that motivates his research: "How can a group use the knowledge and opinions possessed by its members to produce an optimal choice of action for the group as a whole?" The slide is organized into two vertical columns. In the first, the question is posed; in the second, it is illustrated. The photograph is a visual presentation of the dilemma central to Seeley's research: How many hands are up? Must all hands be up before a positive decision is reached? Or only a majority? Or only a plurality? How can the best decision be reached? Below the picture we see the three main stages in group decision-making. The remaining slides in Seeley's presentation, as expressed in its title and subtitle, all contribute to making the case that decision making in human groups and bee swarms is analogous.

In the next slide we reproduce, figure 7.11, Seeley quantifies this decision-making process for honeybees scouting for a new nest. The slide contains a time series of eight graphs, which are read in the same direction as one reads written text. These eight represent vote totals for eleven potential new homes for the hive, taken over a sixteen-hour period; the scouts "vote" by performing a dance in which they "waggle" in the direction of

their choice. The angle of the waggle indicates the direction of a good pros-
pect; the duration of the waggle, its distance.

While the slides that accompany science lectures should form a coherent set, they are not meant to replace the lecturer. In this particular case,
Seeley must tell his audience how to read this graph. He must explain that
the beehive appears as a circle with black dot in the center, that the arrows
indicate the direction and relative distance of a potential new hive, as determined from the waggle dance. He must point out that each arrow has a
letter code indicating a potential home, plus a number indicating how many
scout bees voted in favor of it. The thickness of each arrow, he must point
out, correlates with the number of votes received by a given site. Although
decoding this graph in detail takes a considerable amount of scanning and
matching, with the guidance of the lecturer, its message comes through
clearly because of the ingenuity of the slide's visual design. Site G first appears in the second graph with four votes, a thin pencil pointing southwest. As support for it grows steadily over time, the pencil grows in thickness, until in the eighth graph it has become a single overpowering wedge.
The slide contains another useful visual device. A scale and compass in the
fourth frame at the upper right inform us that the new home (labeled "G")

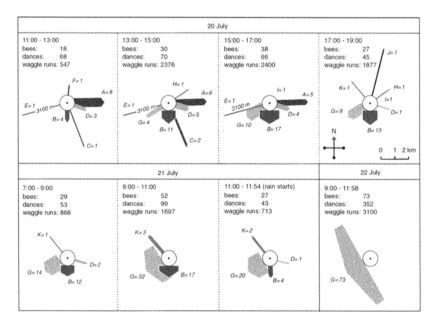

Figure 7.11. Slide graphing data on honeybee decision making for new nest selection.
Votes for site G increase as follows: 0, 4, 10, 9, 14, 32, 20, 73.

Results (note: winner takes all)

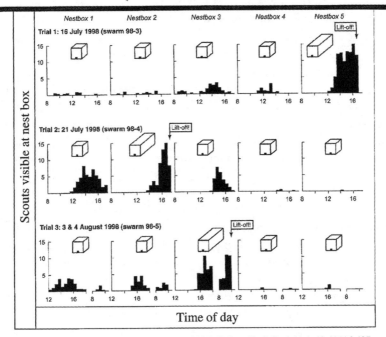

Seeley & Buhrman (2001) Behav. Ecol. Sociobiol. 49:41416-427.

Figure 7.12. Slide graphing data from three trials on honeybee nest selection.

lies about two kilometers to the southwest. Once the scout bees have made their decision, the swarm heads for its new home.

Seeley has invented a visual language appropriate to representing his field data. In this series of graphs and diagrams, we actually see, actually experience, the process of decision making in all of its complexity over time; it terminates in an "aha!" moment represented at last by the utter simplicity of the large wedge of the final panel.[6] This extended exposition is not a distraction because the focus of the audience and the speaker is always the same. (Those who want to study the graph more carefully can consult the journal article cited at the bottom right of the slide.)

But is the decision that has just been illustrated the right one? Does the quorum of scout honeybees always choose the best site? In the set of bar graphs that form the next slide we reproduce, as figure 7.12, Seeley answers this last question.

6. The visual persuasiveness of this slide is much more apparent in the original color slide.

In his lecture, Seeley would have told his audience that, in each of these fifteen graphs, the x-axis represents the time of day; the y-axis, the number of scouts visible at a potential nest box. He would also have explained that the three-dimensional boxes hovering over each graph represent potential nests, the bigger rectangular one being more commodious and therefore superior to the smaller square one.

In Seeley's visual language, in each row, in each trial, lift-off for a new home explodes syntactically: In the set of graphs for each trial, the most commodious "hive," the highest bar, the lift-off box, and the arrow coincide, a climax of visual emphasis. The slide shows that the bees are right every time. But Seeley's research team did more than three trials. And as it turned out, the bees were not right all the time; the slide does not display all of the data. In his lecture, Seeley must have clarified to his audience that the bees were correct only four times out of five. He must also have explained, as he did in his technical report cited at the bottom right of the side, that "this outcome is unlikely to have occurred simply by chance, i.e., if the swarms had made random choices" (Seeley, Visscher, and Passino 2006). The slide itself does not reveal all of the argument; the lecturer reveals it by means of the slide. With the guidance of the lecturer, the main point shines through. Any attendee wanting to investigate further can consult the reference at the bottom right.

The next slide, figure 7.13, reinforces the point Seeley has just made. What is *symbolized* in the graph is *depicted* in the drawing: as the election plays out, the scouts eventually concur and waggle toward the same new home, then the bees emigrate en masse.

All good PowerPoint presentations conclude with one or more slides devoted to the main lessons the lecturer wants the audience to take home. Seeley's conclusion appears in a sequence of three slides. The first acknowledges the well-known problem of "group think" over individual creativity. In contrast, the next slide, figure 7.14, uses a catchy title, numbered list, and photo of a beehive to highlight the benefits of group decision-making. It sums up for its audience the three main lessons that the bees teach us. Supporting each lesson are one or two scientific observations about the scouting bees. This summary slide is a good example of an appropriate use of a numbered list, one of PowerPoint's default conditions.

In a clever turn, when the speaker presses forward with the slide show, the list in figure 7.14 remains as is but the bees in the photo on the upper right disappear. A human brain magically appears in their place, dramatizing the analogy between the swarm and us. This sequence of two slides in one encapsulates Seeley's main point in a nutshell.

Seeley does not tell a story in the sense that he does not relate a series

Dynamics on swarm cluster and at nest sites
that underlie swarm decision making
10:00
13:00
16:00
Seeley, Visscher & Passino (2006) Amer. Scientist 94:220-229

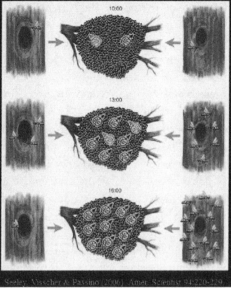

Figure 7.13. Slide depicting honeybee decision making.

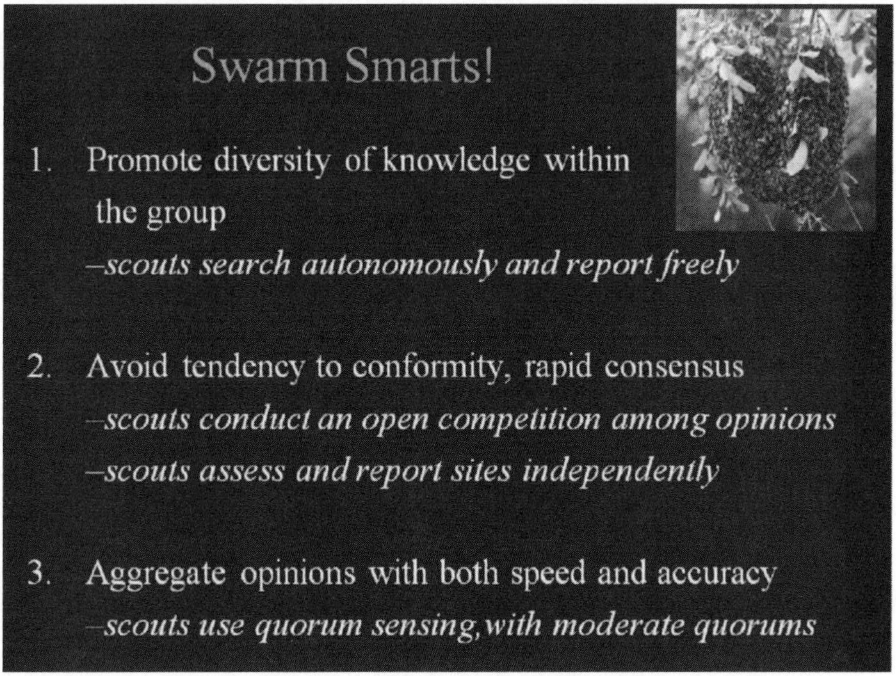

Swarm Smarts!
1. Promote diversity of knowledge within the group
–scouts search autonomously and report freely
2. Avoid tendency to conformity, rapid consensus
–scouts conduct an open competition among opinions
–scouts assess and report sites independently
3. Aggregate opinions with both speed and accuracy
–scouts use quorum sensing, with moderate quorums

Figure 7.14. Summary slide from "House Hunting by Honey Bees."

of unique events that form a coherent whole. Rather, he reveals the details of a process that repeats itself over time: the way in which swarms of bees reach a decision about the location of a new hive. Seeley claims that the bees reach this decision by means of a quorum, one that is far more often right than wrong. The data he amasses and the visuals he displays support this claim, a claim he extends to human groups.

While Seeley's research was time consuming, requiring patient attention over long periods and sedulous attention to detail, in his PowerPoint lecture, his argument has a structure that is simplicity itself. He begins with an introductory slide posing a provocative question in whose answer his audience might be interested. This is followed by a sequence of slides on the basics of bee behavior, revealed while their house hunting simultaneously unfolds. We then see several slide sequences, all revolving around the answers to four specific research questions. His concluding slides summarize and generalize his findings. Throughout, as in Dzurisin's presentation, the crucial unit of analysis has not been the individual slide, but the extent to which the individual slide is integrated into the lecture as a whole and the extent to which the verbal and the visual successfully interact.

IVERSON'S ARGUMENT FOR AN AUDIENCE OF PEERS: SCIENCE AS PRODUCT

Mercier and Sperber have synthesized a provocative argumentative theory. Their conclusion is that reasoning evolved for the "production and evaluation of arguments in communication" (2011, 58). Based upon a wealth of psychological research, the theory has two main tenets. First, while people in general are skilled arguers, they "are not after the truth but after arguments supporting their views" (57). Second, this tendency toward persuasion over truth can be mitigated if the participants in an argumentative exchange exert "epistemic vigilance," that is to say, "share an interest in discovering the right answer" (72). This is what scientists try to do.

For them, crafting arguments is a professional obligation. Their reputation stems from creating innovative knowledge claims about the natural world and arguing for their truthfulness before an audience highly skilled, knowledgeable about the topic, and professionally skeptical. The audience members are also bound by an interest in the "strongest" argument, one surviving any challenges or counterarguments. In the long run, all benefit from this process. The burden of proof on the arguers for a new knowledge claim is high, sometimes requiring years of research and many challenges before acceptance by the discourse community. Since the early twentieth century, the typical medium for expression for these arguments has been the research journal, technical book, or presentation at a professional meeting. In chapter 4, we discussed several examples from research articles.

Here we examine an example PowerPoint presentation delivered at a meeting of geologists. In contrast to Dzurisin's and Seeley's implied audiences, Richard Iverson's "Dynamic Model of Seismogenic Volcanic Extrusion, Mount St. Helens, 2004–2005" largely excludes the general public. This is because of its argumentative structure, forced on Iverson by the need to support the claims he makes to an audience of peers. While complex arguments are difficult to follow in print, readers can underline, take notes, move back and forth from words to figures and tables, and, of course, reread. The audience of a PowerPoint presentation has no such recourse: it must view one slide after another at a relatively uniform pace.

A comparison between Iverson's PowerPoint lecture and an accompanying article in *Nature* (2006) underlines the expository constraints oral presentations impose. In the lecture, the three facts essential to building the model are highlighted in individual slides that make their relationship to the model clear; in contrast, in the article in *Nature*, these facts are incorporated in a section entitled "Eruptive behavior and seismisticty," embedded in paragraphs that do not identify their important role. For example, Fact 2 is presented as follows:

> The newly extruded dacite formed a series of spines with freshly exposed surfaces coated with fault gouge [the crushed and ground-up rock produced by friction between the two sides when a fault moves]. The most prominent spine emerged in winter of 2005 and had a remarkably smooth, symmetrical whaleback form (Fig. 1 and Supplementary Movies 1 and 2), but most of the spines were partly obscured as they pushed past glacial ice and previously extruded rock. (Iverson et al. 2006, 439)

There is an additional obstacle to the full understanding of this paragraph: seeing for oneself what is described involves, first, turning to the following page to view the related photographs and graphs, then going online to view two short movies of the eruption's effects. These radical shifts create spatial discontinuities between the verbal and visual, making their interaction difficult. This is an obstacle that the more visually dominant PowerPoint presentation can easily circumvent.

The comparison between written and oral scientific communication, between Iverson's *Nature* article and his PowerPoint lecture, shows that the close integration of the verbal and the visual is vital for maintaining audience attention and comprehension in professional oral presentations where argument is the dominant mode. Iverson's presentation exemplifies this principle. The opening sequence centers upon three crucial facts known to volcanologists as a result of the close study of seismic activity at Mount St. Helens, especially during its most recent eruption:

Fact 1. A large plug of volcanic rock formed, fitting into the volcano cone like a loose-fitting cork in a bottle. Because of the powerful turbulence below, for about a year this plug had moved up and down at a nearly constant rate.

Fact 2. The newly formed extrusions created a series of spines whose freshly exposed surfaces exhibited striated fault gouge formed by the movement of the plug against the sides of the volcanic cone. Analysis showed that the rate of movement of the plug weakened over time.

Fact 3. In the area of the volcano, repetitive "drumbeat" earthquakes caused by the oscillation of the plug occurred about every hundred seconds, had a magnitude of less than two, and were centered at depths of less than a kilometer, directly beneath the plug the eruption created.

All three facts appear as slide headlines and are accompanied by a photograph or graph that makes the verbal visual. From these facts, Iverson constructs a mechanical model of the eruption. From this mechanical model, he infers its mathematical counterpart. From this counterpart, he deduces a series of empirical consequences, one subset of which, as it happens, closely matches the facts on the ground. It is this coincidence that corroborates the plausibility of the mechanical model; it is this simple chain of reasoning that constitutes his technical argument.

In our discussion of Iverson's strategies of presentation, we will focus on the sequence that derives from Fact 2. In figure 7.15, the technical language in the text box clearly assumes a knowledgeable audience, one that realizes that "rate-weakening frictional strength" is not an oxymoron. This audience is seeing with their own eyes what soon will be captured by a mechanical model. First, they see the photograph without the caption; then the caption enters from the left; at the same time, the arrow points directly to the feature in question. In this slide, the photograph converts the verbal expression "striated fault gouge that coats the surface of the newly extruded dactite plug" into its visual counterpart.

The next slide, figure 7.16, gives visual expression to the predicate of Fact 2: "exhibits rate-weakening frictional strength." The presence of the same heading in the same format binds these two slides into a single

Figure 7.15. Slide picturing striated fault gouge as it exists in nature.

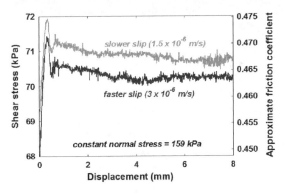

Figure 7.16. Slide showing the forces behind the geophysics that created the striated fault gouge in figure 7.15.

cognitive unit. The graph in figure 7.16 indicates that slip, the upward motion of the volcanic plug, increases as a consequence of reduced friction. In
the graph, shear stress, stress parallel to the face of the material, is measured
in kilopascals (kPa), ten of which equal one atmosphere of pressure. The
graph shows us two curves: an upper red one for a slower slip, and a lower
blue one for its faster variety. In the slide, the different colors make discrim-
ination and comparison easier. The two slides (figs. 7.15 and 7.16) combine
to give us a multifaceted picture of the fact expressed in the single heading:
we first see the gouge as it exists in nature; we next see its transformation
into standard physics measurements in support of *Fact 2*. The progression is
from iconic images of the lava dome to indexical graphs depicting the cause
behind the effect, a calculable nexus of forces.

   To transition from the established facts to the next stage in his argument, Iverson presents a brief movie. Shot from near the top of Mount St.
Helens looking into the crater, it shows the undulations in dome growth
over a year's time. The next slide, figure 7.17, depicts a physical model of
what the audience has just witnessed through the magic of time-lapse video.
In the slide, Iverson views this volcanic system as a damped oscillator, one
whose back-and-forth movements are reduced over time. These movements
are repetitive displacements ("slip") that are intermittent ("stick"). It is this
start-and-stop motion that causes the "drum-beat" earthquakes mentioned
in Fact 3. Figure 7.17 models the mechanism behind the spatiotemporal
magnitudes of motion. A plug of mass m under pressure from the rising
molten rock or magma P, and counteracted by the force of friction F, os-
cillates around a point of equilibrium as a consequence of the drum-beat

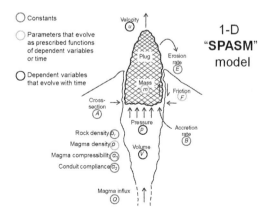

Figure 7.17. Slide showing physical model of extrusion dynamics. "SPASM" is the
model's name, standing for "Seismogenic Play of Ascending, Solidifying Magma."

If $\kappa = 0$, $B = Q$, and t_0 is constant, behavior of numerical solutions depends almost entirely on D evaluated at the equilibrium slip rate $u = u_0 = Q/A$:

$$D = \frac{1}{2}\frac{t_0}{m_0}\left[\frac{dF}{du}\right]_{u=u_0} = \frac{1}{2}\left(c\lambda\mu_0\frac{gt_0}{u_0}\frac{u_0}{u_{ref}}\left[1+\left(\frac{u_0}{u_{ref}}\right)^2\right]^{-1/2}\right)$$

which simplifies to

$$D \approx \frac{1}{2}\left(c\lambda\mu_0\frac{gt_0}{u_0}\right) \qquad \text{if } u_0/u_{ref} \gg 1$$

Figure 7.18. Slide showing mathematics behind the physical model.

earthquakes that diminish in intensity over time (see fig. 7.9). Beneath each variable, parameter, and constant in the diagram, their circled shorthand symbols appear. In the slide, these are color coded for ease of identification: brown for constants, blue for parameters, and red for variables. The color mapping and symbols link the mechanical model to the mathematical equations that follow in the next slides. Comprehending this diagram in full takes considerable scanning and matching among the semiotic components; however, lecturers can rapidly direct the audience's attention to the most salient visual points.

The remainder of Iverson's lecture concerns the construction of the mathematical counterpart of this mechanical model, and the derivation of its empirical consequences. In our discussion, we focus on two related slides in the mathematical model sequence. The slide reproduced next, figure 7.18, would not sit well with a general audience. But professionals attending a conference presentation about a new model of volcano extrusion dynamics would be unfazed by such conventional mathematics. This set of equations is embedded in a single sentence that conveys a single thought: when the plug velocity is high, the damping factor (D) is a function of the rate-weakening strength (c) multiplied by that velocity. Moreover, this equation links the mathematical model to its mechanical counterpart: m in the equation matches m in the model, u matches u, and F matches F.

The final Iverson slide we reproduce, figure 7.19, contains two graphs, each conveying a different facet of the same concept: as the magnitude of D decreases, the eruption becomes more violent. The graph on the left shows the shape of the stick-slip cycle over time; the graph on the right shows the relationship of this cycle to the pressure and velocity. In viewing this slide, the audience first sees only the pair of graphs and their labels; at this

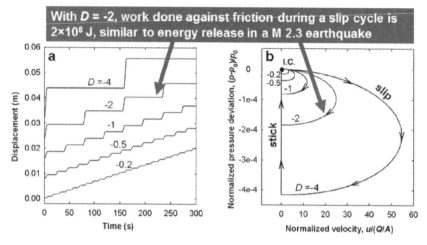

Figure 7.19. Slide plotting D (dimensionless damping) calculated from Iverson's mathematical model of the Mount St. Helens eruption.

juncture, a blue bar moves from the left across the screen, its twin arrows pointing to the relevant curves for $D = -2$, the condition best approximating that of the actual Mount St. Helens eruption. While full comprehension of these twin graphs would in all likelihood involve considerable cognitive effort, even for a volcanologist, the blue bar directs the audience's attention to the most salient point. Further, it is this blue bar that creates emphasis throughout Iverson's presentation.

Iverson's presentation ends as do most for professional audiences: with a bulleted or numbered list of conclusions. Most such presentations also start with an introductory list that establishes a research problem. Because this presentation does not have such a list, we have taken the liberty of creating this "effective redundancy," an essential component of good presentations (Doumont 2002), by borrowing from the introduction to the *Nature* article by Iverson and his coworkers:

- Volcanic eruptions are difficult to model.
- The recent eruption at Mount St. Helens exhibited near-equilibrium behavior for about a year.
- This behavior includes nearly steady extrusion of volcanic material coupled with periodic shallow "drum-beat" earthquakes.

- This behavior will be modeled mechanically and mathematically as a damped oscillator.

We have no doubt that Iverson's argument would be mostly unintelligible to general audiences. Yet, despite the complexity of the science, the logic of Iverson's argument is straightforward—a shift from the facts on the ground to a mechanical model, followed by a deductive movement from the mathematical counterpart of this model to a series of empirical consequences, one subset of which corroborates the model by approximating the actual conditions of the Mount St. Helens eruption. Iverson's is an argument adapted to the interests and abilities of a professional audience sharing "an interest in the right answer" (Mercier and Sperber 2011) to a complex problem. Throughout, as with Dzurisin and Seeley, the crucial unit of analysis has not been the individual slide; it has been the extent to which the individual slide is integrated into the presentation as a whole and the extent to which this integration promotes verbal-visual interaction in the creation of meaning.

LOSOS'S ARGUMENT FOR AN AUDIENCE OF PEERS: SCIENCE AS PROCESS

At first glance, Losos's representation of science exactly parallels Iverson's. In "Integrating Ecological and Evolutionary Studies of Biological Diversity: Patterns and Processes of Adaptive Radiation in Caribbean Lizards," he summarizes his claim to new knowledge, the product of a decade of research concerning the adaptive radiation of ecomorphs of the genus *Anolis*: "On each island of the Greater Antilles (Cuba, Hispaniola, Jamaica, and Puerto Rico)," he states, "anoles have radiated for the most part independently, nonetheless producing essentially the same set of habitat specialists, termed 'ecomorphs.'. . . Members of each ecomorph class are species that are very similar in morphology, ecology, and behavior despite their independent evolutionary origins" (Losos 2010, 628). In another lecture, one accepting the 2009 E. O. Wilson award for midcareer achievement and published in *American Naturalist*, Losos states his broad theoretical aim, the clarification of an important concept in evolutionary biology:

In the first part [of this lecture], I suggest that the common idea that "ecological opportunity" triggers adaptive radiation is usually correct, but that ecological opportunity is neither necessary nor sufficient for radiation to occur: theory and methods remain to be developed before we can fully understand why radiation occurs at some times and not at others. In the second [part of this lecture], I focus on a specific aspect of adaptive radiation, the extent to which the course it takes is predictable and deterministic. (Losos 2010, 624)

In this award lecture, Losos's images are subordinate to words: there are only two photographs, two cladograms, and one hierarchical flowchart. In contrast, in his PowerPoint lecture, Losos explores these same themes in ninety-four slides in which words are invariably subordinate to images. The degree of this subordination is made clear by comparing the presentation with the articles cited as its sources. There are nine articles containing approximately 80,000 words and exactly 124 images: about one image for every 640 words. In contrast, in Losos's PowerPoint there are approximately 750 words and exactly 120 images: about one image for every 6 words. Of course, this hundredfold difference exaggerates the discrepancy between words and images: it ignores the contribution of the lecturer. Still, there are ninety-four slides: assuming a presentation of about an hour, on the average each slide could be on view for only thirty-eight seconds, just enough time to make a point or two about the image displayed before moving on, just enough time to rely on verbal-visual interaction to convey meaning.

To trace the adaptive radiation of one genus of lizard, Losos shifts from one visual representation to another. We progress from a tree diagram mapping spatiotemporal relationships to a graph representing data trends to a photograph allowing us to see the object of study. Figure 7.20 reproduces a tree diagram ("cladogram") of the ancestry of the trunk-crown anole. The words on the slide are minimal: the names of the four Caribbean islands in the study and the name of the lizard type. Although understanding the details of this tree diagram requires some training, the main point is clear once you know that time flows from bottom to top, and that the lines represent the evolution of trunk anoles in the context of the evolution of every known species of *Anolis*. Reading from left to right in the diagram, those from Cuba are emphasized in red, those from Puerto Rico in violet, those from Jamaica in green, those from Hispaniola in brown. In the words of an article by Losos and a collaborator, "On each island, evolutionary diversification has proceeded for the most part independently, producing on each island the same set of habitat specialists adapted to use different parts of the vegetation" (Losos and Ricklefs 2009, 833–34). Figure 7.20 is indexical: it represents adaptive radiation, the evolutionary force behind lizard speciation.

Figure 7.21 focuses on the evolutionary origin of the genus, placing two fossil *Anolis* specimens, represented by plus signs, in a context that graphs

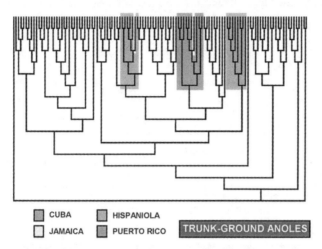

Figure 7.20. Tree diagram of the trunk-crown anoles that evolved independently on Cuba, Jamaica, Hispaniola, and Puerto Rico.

the correlation between tibia and snout vent length of five contemporary ecomorphs, physical indications of adaptation to different environmental conditions. The visual alignment of plus signs with solid triangles provides compelling evidence for a strong correlation between the fossils and the present-day *chlorocynus* or the trunk-crown anole—a species of lizard ecologically specialized for life high on the trunks and in the crowns of trees. Figure 7.22 ends this sequence with a photograph of one of the fossil lizard specimens, trapped in amber. The photograph also includes a millimeter scale because pictures can represent only relative, not actual size. These fossils have now been identified as evidence for the "ancient diversification into ecomorph niches," a phrase from a bulleted list seven slides on.

Figures 7.20 to 7.22 form an integrated cognitive and visual unit. Figure 7.20 is indexical; it represents the evolutionary process in the form of adaptive radiation, the force behind lizard speciation. Figure 7.21 is indexical as well; it represents the relationship between two fossils and their contemporary progeny. Figure 7.22 is iconic, a photograph of a lizard trapped in amber; it is also symbolic, meant to represent all such lizards with a similar ratio of tibia to snout-vent length. But placed in the context of figures 7.20 and 7.21, it is also indexical: it counts as evidence for adaptive radiation over evolutionary time.

For Jonathan Losos, as for Iverson, PowerPoint means representing science as a product in which the link between the iconic and the indexical is visually central; for Losos, though not for Iverson, PowerPoint is also a

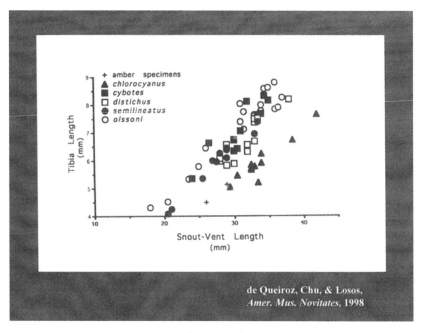

Figure 7.21. Scattergram comparing data trends for tibia versus snout-vent length of five existing Hispaniolan species and two fossils, one of which is represented in figure 7.22. The five Hispaniolan species are *A. chlorocyanus* (member of the trunk-crown ecomorph class), *A. cybotes* (trunk-ground), *A. distichus* (trunk), and *A. olssoni* and *A. semilineatus* (both grass-brush).

Figure 7.22. Slide with ruler showing lizard fossilized in amber.

Figure 7.23. Slide showing Jonathan Losos on field site in the Bahamas.

means for re-creating for one's audience the world in which this shift can be successfully realized. Losos harnesses nature to his purposes, first, by intervening in it, permanently changing its course in the interest of science; second, by reimagining it, turning a sequence of natural events into a controlled experiment.

Figure 7.23 is a snapshot of Losos on a small, deserted island in the Bahamas. It is a very sunny day, and, very sensibly, he is wearing sunglasses, a broad-brimmed hat, a long-sleeved shirt, long khaki pants, and comfortable shoes. What has this snapshot, a photograph that would not likely appear in a scientific journal, to do with Losos's science? With intervening irreversibly in the course of nature? A great deal. As a snapshot taken on a vacation in the Bahamas—with Losos positioned off to the far side of the frame—this photograph would be a good candidate for the family album. But Losos is posing on one of twelve small islands, a cluster being transformed into an experimental site, one designed to change the course of nature in order to reveal one of its effects. On six control islands, a lizard ecomorph, *Anolis sagrei*, will run free. On six experimental islands, a predatory lizard, *L. carinatus*, will be introduced, a creature for which *Anolis* is a prey. Losos's experiment will test whether natural selection operates as a consequence of predation. In the article that reports their results, Losos and his coworkers confirm their hypothesis: "the presence of the predatory *L. carinatus* favoured longer-legged male lizards, which can run faster, and favoured larger females, which are both faster and harder to subdue and ingest" (Losos, Schoener, and Spiller 2004, 507). Now that we understand Losos's intent, we are in a position to understand this snapshot as a representation of science as a practice.

Figure 7.24, also from Losos's PowerPoint lecture, is a routine weather map showing the position and progress of Hurricane Frances on September 3, 2004, at 11:46 AM (EST). What has this map, an image that also would not likely appear in a scientific journal, to do with reimagining nature as a collaborator in his endeavors? A great deal. On October 19, 1996, Hurricane Lili struck the Bahamas with a storm surge of five meters, guaranteed to inundate many islands while lizards were still reproducing. On September 14, 1999, Hurricane Floyd produced a storm with a surge of three meters, still powerful and still during the reproductive season, though earlier. Although Lili wiped out populations of *A. sagrei* on the observed islands, Floyd did not. This twofold contrast—of dates and intensity—raised a question: was extermination

Figure 7.24. Weather map depicting Hurricane Frances, just north of Cuba, as it moves toward the American east coast.

the result of the strength of the hurricanes or their timing? What was needed was a missing component in a controlled experiment, a hurricane of low power that arose after the reproductive season was nearly over:

> Obligingly, such a hurricane has now happened. On November 5, 2001, Hurricane Michelle passed over the same islands as did Floyd but generated a storm surge only somewhat > 1 m. Was Michelle unable to affect *A. sagrei* populations much because of its relatively low power, or was Michelle very devastating, just as was Lili, because it was so late in the season despite its lesser power? (Schoener et al. 2004, 177)

By reimagining a sequence of destructive hurricanes as a controlled experiment, including Hurricanes Frances and Jeanne in 2004, Losos has harnessed the weather to his purposes: "we discovered that lizards of the species *Anolis sagrei* on small islands are especially vulnerable to inundation from the storm surge of a hurricane, but they are sometimes able to survive this destructive force as eggs" (Schoener et al. 2004, 177). Timing and intensity both mattered. In Losos's PowerPoint, a weather map originally designed to warn of oncoming danger has been repurposed in the interests of a science other than meteorology.

We asserted earlier that while narrative was the preferred structure of general audience presentations, in professional settings argument was dominant. Losos's PowerPoint complicates this assertion. Certainly, he is making an argument, and certainly that argument is dominant. By means of investigations in the laboratory and the field, a field turned into a laboratory, he determines the ecological and environmental forces that lead to adaptive radiation. For example, by situating an Anole fossil in taxonomic space, he demonstrates the ancient origin of these ecomorphs. But because evolutionary biology is a historical science, Losos is at the same time telling the story of Anole evolution, a unique sequence of events generated by uniform laws. Indeed, he contributes to that story: in demonstrating the role of predation in natural selection, he permanently alters the ecology of six small Bahaman islands. He also alters a story that meteorology tells, a tale of a closely spaced sequence of destructive tropical hurricanes. By reimagining this sequence as a controlled experiment in biology, he turns a story about the weather into a story about evolution.

At the center of Losos's PowerPoint lecture, we have not only science, but also a scientist creating science. Over the centuries, the methods section of experimental articles has grown in communicative importance as reflected in its elaborate technical details (Gross, Harmon, and Reidy 2002,

191–93). But there were few specifics concerning time, place, or persons. Losos reverses this centuries-old trend. Time, place, and persons are back in the picture. We see a date-stamped weather map; we see the islands and the lizards that inhabit them; we see the researchers at work. By means of verbal-visual interaction, Losos tells the story of his science as a practice.

CONCLUSION

In the early days of printing, images could be widely reproduced, but only from woodcuts, some as sophisticated as those of Vesalius, most very crude. Moreover, while the printing of images had been mechanized, their creation in a form suitable for reproduction was still a labor-intensive craft skill. Although the introduction of engraving greatly improved image accuracy, the craft origin of images had still not been superseded. Furthermore, because of technical limitations in book production, engravings were regularly gathered in single plates at the back of books and journals, creating an obstacle for the integration of text and images. Even though the invention of photography mechanized image reproduction, until the end of the nineteenth century technical limitations prohibited mass reproduction in print. Until the end of the twentieth century, moreover, color was rare, also for technical reasons; its reproduction remains expensive to this day. Nevertheless, a glance at almost any page of *Science* or *Lancet* reveals a visual landscape of which people of earlier centuries could only dream: the images on display exemplify the full exploitation of the printed page, abetted by the technological advances of the past two centuries. Even now, however, text dominates; in effect, images form islands in a sea of print.

Throughout earlier centuries, the scientific lecture was held hostage to similar visual limitations, aided only by cumbersome slide carousels and overhead projectors. With PowerPoint, we enter a new realm of communicative flexibility; we have a widely used software program capable of projecting both still and moving images in color. Unfortunately, PowerPoint's default conditions—its six levels of hierarchy for bulleted lists, embedded templates, and many options for slide transitions—carry over into the twenty-first century a culture of print ready to entrap the unwary. In the worst presentations, verbal clutter turns audience goodwill into bewilderment and hostility.

But PowerPoint is a transformative tool in the hands of skilled communicators like Dzurisin, Seeley, Iverson, and Losos. Their presentations focus audience attention on the visual heart of science. As a consequence, their lectures can approach the Heideggerian communicative ideal: "a communion simultaneously with others and with the objects of mutual concern: when we are explicitly hearing the discourse of another, we proximately understand what is said, or—to put it more exactly—we are already with him, in advance, along side whatever it is the discourse is about" (Heidegger 1972, 164; 1962a, 207; translation corrected). This Heidggerian formulation makes it clear that the creative freedom PowerPoint bestows comes with a caveat: never to permit audience attention to stray from the visual center; always promote verbal-visual interaction. PowerPoint presenters also need to be vigilant in providing their audiences with an easily perceived over-arching structure that facilitates this focus, narrative in the case of general audiences, argumentative in the case of audiences of students, college teachers, and peers. In the hands of skilled practitioners, the scientific lecture can be made arresting and memorable. In the hands of someone as inventive as Losos, even the lecture's genre can be reinvented in the service of a somewhat different representation of science.

Despite its considerable communicative potential, however, PowerPoint remains a conservative medium, tied to a past of conventional public presentations. For true innovation in the communication of science, we must turn to the Internet, the subject of our final chapter.

CHAPTER EIGHT

Weaving the Web of Scientific Knowledge: Visuals and the Internet

Socrates: Then he [a person of real knowledge on a subject] will not, when he's in earnest, resort to a written form and inscribe his seeds in water, and in inky water at that; he will not sow them with a pen, using words which are unable either to argue in their own defense when attacked or to fulfill the role of a teacher in presenting the truth. . . . In this regard, far more noble and splendid is the serious pursuit of the dialectician, who finds a congenial soul and then proceeds with true knowledge to plant and sow in it words which are able to help themselves and help him who planted them; words which will not be unproductive, for they can transmit their seed to other natures and cause the growth of fresh words in them, providing an eternal existence for their seed; words which bring their possessor to the highest degree of happiness possible for a human being to attain.

—Plato, *Phaedrus* (360 BCE)

Electronic journals should not and will not be mere clones of paper journals, ghosts in another medium.

—Stevan Harnad, *Public-Access Computer Systems Review* (1991)

From the fourth century on, the codex reigned supreme in the world of publishing, replacing the outmoded scroll. It was more compact, could be inscribed on both sides, and could be more easily stored. But more important to scholars and scientists was its increased usefulness as a tool to facilitate intellectual exchange and advance knowledge. The codex opened the door to such innovations as tables of contents, section headings, page numbers, and indexes. For the first time in the history of intellectual work, researchers could easily look something up. Systems of citation, the fingerprint of the communal activity that is science and scholarship, became feasible. Later, with the conversion from wine press to printing press, the invention of movable type, of inexpensive, durable paper from wood pulp, and, finally, of better and better means of reproducing images, the codex reached its current state of maturity. In the past four hundred years of its development, in the form of the book and then the research journal, it has been at the center of scientific communication.

Nevertheless, as the optimal medium of scientific communication, the codex has nearly run its course. Impelled by the Internet,[1] currently the principal means for first reporting original scientific discoveries, the scientific article is now undergoing a transformation into a very different future. This prediction can be made with some confidence because, in a real sense, that future is already here. The transition of the scientific literature from printed pages we turn to screens we scroll and click on is already well under way. In fact, most significant scientific journals now offer electronic and print versions of current issues, in addition to a substantial electronic archive of back issues. Some, such as those published by the Public Library of Science, appear only in electronic format. Economics dictates that all duplication will eventually come to an end. It is only a matter of time before the journal in print moves onto the endangered species list.

In *Electronic Literature* (2008), N. Katherine Hayles notes that "at first [this literature] strongly resembled print and only gradually began to develop characteristics specific to the digital medium, emphasizing effects that could not be achieved in print" (59). In this our final chapter, we discuss some important communicative differences between the Internet and printed articles in the scientific literature.[2] In effect, the Internet has reinvented the scientific article and related communications. First, the Internet has made possible a new form of Socratic dialectic, a sense that intellectual interchange may now come close to what Socrates imagined—peer-to-peer dialogue differing from conversation only in its rigor. Second, the Internet has fostered a new form of collaboration—one indifferent to whether its participants are persons, networks of concepts and inferences, or computer programs managing those networks. Third, the Internet has fostered the

---

1. We use the broad term "Internet" to encompass the World Wide Web, e-mail, bulletin boards, blogs, and so on, though normally we are referring to the World Wide Web.

2. Most of our examples come from *Science, Nature*, and *New England Journal of Medicine* not because they are and have been for about a century the most prestigious journals in science and medicine, and not because they are representative of the scientific Internet literature at large, but because they are in the avant-garde. With a large subscription base plus advertising, they have the funds and other required resources to lead the way into the unchartered post-Gutenberg seas.

creation and enabled the dissemination of dynamic visual representations such as movies, 3-D rotatable representations, and interactive graphics.[3] One of the important consequences of this enhanced palette has been the emergence of a new form of virtual witnessing: for the first time, the reality that readers experience can very nearly coincide with field and laboratory experience. It is not simply that creators and consumers see the world as a calculable nexus of forces operating on spatiotemporal magnitudes of motion; this has always been the case in modern science. It is that the consumers can reexperience what the creators have already experienced.

We would not do justice to scientific communication on the Internet if we limited our analysis only to interactions among researchers. To do so would be unreasonably to restrict ourselves to a limited segment of—to borrow a famous phrase from Roland Barthes in a different context—the "festival of affects" available through the Internet. Accordingly, in the latter part of this chapter we focus on medical communications designed for practitioners and interested citizens. In medicine, the division between a small cohort of researchers and a far larger one of clinicians and patients has created a need not only for sharing among peers, but for informing and instructing broader audiences. For example, the Internet sites of medical journals now contain short movies designed to inform physicians and others of new clinically relevant advances. We look at one of these. While the system of delivery is new, these videos largely conform to a well-established genre: the news story. We also examine two videotaped instances of medical instruction: the modeling of diagnostic protocols and the integration of preferred medical procedures into existing practices. In both cases, established communicative genres are also transformed by the new delivery system: in the first, the grand rounds; in the second, the instructional manual.

In devoting a chapter to Internet science, we face two challenges, both created by the technology we propose to analyze. The first is the product of the very innovations that are our focus. Within the confines of a book, a physical object made of paper, ink, twine, and glue, we can describe, but we cannot adequately illustrate the innovations that are our subject. This chapter contains no images in color, no video productions, no interactive graphics, no 3-D chemical structures, no sound bites. Nothing could demonstrate the limitations of print and the advantages of the Internet better than our in-

---

3. Such visuals have been used long before the Internet, but they are now appearing for the first time in what had previously been print-only media such as research articles and books, where they are being used in innovative ways to construct arguments and narratives. This is true for both the humanities and sciences.

ability to reproduce these still and moving images, these sights and sounds, so central to current science. But we take comfort in the fact that readers can compensate for this deficiency; they can turn to the Internet sites in which these images appear. Even those who do not have access to the currently restricted domains of such journals as *Science* and *Nature* are not without resources, thanks to the growing open access movement. Especially significant at this time is the cluster of journals published by the Public Library of Science. These are currently the best illustration of the fact that, for the discerning searcher of the Internet, the future is already here.

There is a second technological challenge, the allegation that by the time this book is in print, everything we say about science and the Internet will be seriously out of date. In one sense, we agree: we certainly have no way of anticipating developments that depend on the creativity and ingenuity of scientists, journal editors, and learned societies. Still, we think that our analysis will remain useful. We are not offering a tour of the seven wonders of Internet science soon to be replaced by seven new wonders; we are providing a framework for the understanding of what the Internet can mean for scientific communication, a framework grounded in the history of its oral and written communication.

THE INTERNET AND REVIVAL OF PLATONIC DIALECTIC

The Internet is not without critics. Nicholas Carr credits Martin Heidegger with foreseeing, in *The Question Concerning Technology*, that "the frenziedness of technology may entrench itself everywhere to such an extent" that it brings to an end the sustained "meditative thinking" required to grapple with truly complex ideas that are not solvable by "calculative thinking" (Carr 2010, 222). Such critics view the Internet as a giant step in the direction of the dehumanization of communication, a shrinking of human potential in the guise of expansion. Others have expressed concern that the level playing field the Internet provides for publication will open the floodgate to third-rate science masquerading as the real thing, and that first-rate science will remain buried in the resulting flood. We argue that this uneasiness has its roots in the trepidation attendant on any radical alteration in the medium of intellectual exchange. The shift from speech to writing in ancient Greece is our historical example; the shift from print to the Internet, we claim, is analogous.

Like us, Plato lived in a time of media transition. In both the *Phaedrus* and the *Seventh Letter*, he reflected on and deplored the deleterious impact of this transition. Plato's was the classic rationalization of the anxiety at-

tendant on a major shift in the means of scholarly and scientific communication. There was the loss of an important skill: the ability to memorize. In
a time when only actors memorized, cultural resources would no longer be
distributed among minds. Instead, all would rely on a prosthetic device: the
written word, stored in scrolls. Writing had a consequence even more dire: it
undermined the very source of intellectual advance, the dialectic process in
which mind challenges mind in the interest of truth. In the *Seventh Letter*
Plato says of those so handicapped by the culture of writing that

> none . . . will ever learn to the full the truth about virtue and vice. For
> both must be learnt together; and together also must be learnt, by com-
> plete and long continued study, as I said at the beginning, the true and
> the false about all that has real being. After much effort, as names, defi-
> nitions, sights, and other data of sense are brought into contact and fric-
> tion one with another, in the course of scrutiny and kindly testing by
> men who proceed by question and answer without ill will, with a sudden
> flash there shines forth understanding about every problem, and an intel-
> ligence whose efforts reach the furthest limits of human powers. There-
> fore every man of worth, when dealing with matters of worth, will be far
> from exposing them to ill feeling and misunderstanding among men by
> committing them to writing. In one word, then, it may be known from
> this that, if one sees written treatises composed by anyone, either the
> laws of a lawgiver, or in any other form whatever, these are not for that
> man the things of most worth, if he is a man of worth, but that his trea-
> sures are laid up in the fairest spot that he possesses. But if these things
> were worked at by him as things of real worth, and committed to writing,
> then surely, not gods, but men "have themselves bereft him of his wits."
> (360 BCE, Stephanus no. 344b–c; quotation is from the *Iliad*, 7.360)

These remain powerful arguments. Print, the now-pervasive offspring
of the written word, does impede dialectical exchange. Part of the impedi-
ment is due to the time gap between the creation of socially certified knowledge in the form of a manuscript that has passed through some sort of peer
review and its dissemination in print. In the pre-Internet era, books have
typically waited as much as two years from final revision to publication; articles, up to a year, a constraint imposed mainly by the mechanics of printing and distribution. In fast-moving fields, the "new" information may be
ancient history by official publication day.

While Plato may see clearly the losses attendant on the shift from
speech to writing as a medium of intellectual exchange, his is hardly a bal-

anced view. As Alexander Nehamas and Paul Woodruff (1995) point out, arguments like Plato's reappear whenever a new medium of scholarly and scientific exchange appears on the horizon:

> We generally tend to connect the older medium to rationality and to suc-
> cessful communication exclusively; we tend to describe the new one as
> less rational and much less likely to succeed in communicating ideas.
> What is often true in such discussions is that we make an unfair com-
> parison: we judge the new medium according to its ability to communi-
> cate the type of ideas for which the older medium has been designed, and
> it is no surprise that it fails in that regard. Moreover, we tend to identify
> the ideas suited to the old medium and the manner in which that me-
> dium communicates them with what is rational. Accordingly, even if
> the new medium is sometimes judged to be successful in communicat-
> ing its own ideas, we are tempted to consider these ideas at best as infe-
> rior to the former, at worst as irrational and harmful. (xxxvi)

Nehamas and Woodruff point out that writing does in fact have a significant advantage over speech: it is better able to express complex ideas and arguments. It is difficult to image Kant's first critique or Einstein's arguments in favor of special relativity issuing from any exchange that was exclusively spoken.

This example suggests a wider claim: that a culture-wide change in medium is also an alteration in our fundamental intellectual habits, a change in the established set of presuppositions that affect the kind and content of communications. We think this is what Marshall McLuhan had in mind when he said that "the medium is the message." While it is difficult to extract from McLuhan's voluminous writings exactly what he meant by his famous slogan, one possibility is particularly germane to the Internet, a medium that incorporates other media. Like the orchestra conductor putting her personal stamp on the individual instrumental and vocal voices under her power, the Internet transforms the media it incorporates—television, movies, print, radio. In so doing, it creates for the first time since the invention of writing a future that can reap the benefits of Platonic dialectic while, at the same time, doing full justice to the complexities of the written word and scientific visual.[4]

---

4. Of course, the Internet as currently operated is not always a Platonic forum for civil debate. There is a dark side in the form of inflammatory and offensive speech and viral spread of misinformation. See Levmore and Nussbaum 2011. As we assess the present situation, this is a serious problem for the Internet in general, but not as it applies to scientific communication.

PLATONIC DIALECTIC AND INTERNET VISUALIZATION

The dialectic potential of the Internet is already being realized with scientific blogs, and not surprisingly, verbal-visual interaction is as important here as in other forms of scientific communication; indeed, it may be more important. Consider the case of an article prepublished in 2009 and later withdrawn from the prestigious *Journal of the American Chemical Society* (*JACS*). Entitled "Reductive and Transition-Metal-Free Oxidation of Secondary Alcohols by Sodium Hydride" (Xinbo Wang, Bo Zhang, and David Zhigang Wang), it claimed a counterintuitive result: sodium hydride, a strong reducing agent, had effectively oxidized an alcohol and produced a corresponding ketone. The paper had passed peer review; however, its claims did not survive the review of other peers, those outside the chosen circle of *JACS* reviewers. Postpublication, the blogosphere erupted in a vigorous debate. Two respondents—Paul Docherty and Jean-Claude Bradley—refuted the paper's claims, not by further textual scrutiny, as in peer review, but by replication: they attempted to obtain the paper's results using its methods. On his blog, *Totally Synthetic*, Docherty recounted his failure to do so in an engaging informal style:

> As many of you will have noticed in the comments to the previous post (which was thoroughly hi-jacked), an intriguing paper has been published in JACS by Xinbo Wang, Bo Zhang and David Zhigang Wang. In this, they suggest it is possible to oxidise benzylic alcohols to the corresponding ketones using sodium hydride (amongst other chemistry). Given that sodium hydride is, well, a hydride—this is quite something. Does it work? Hard to say without giving it a go, so I am [THF stands for tetrahydrofuran, and is a common ether-type solvent].

> We had this gear in the lab, so I'm giving it a go. (Docherty 2009)

The visual in this blog posting is a map of the spatial arrangements and bonding of elements in a benzylic alcohol (left side) and a ketone (right side), a molecule formed by reaction of the former with sodium hydride (NaH) and tetrahydrofuran (THF). It is provisionally indexical. Both the verbal and visual ask the same questions: Is this strange reaction possible? Does the structure on the left combine with the compounds flanking the arrow to yield the structure on the right? And from the language in the text and the large question mark in the middle of the visual, it would appear that Docherty did not believe so even before he assembled the necessary gear and gave "it a go."

Following this introduction comes a description of Docherty's step-by-step synthesis method interspersed with photographs of key steps, such as a beaker with thermometer in solution. As a result, fellow chemists can actually see the color and consistency of the prepared solution, size and shape of the beaker, and placement of the thermometer—useful images often left out of print journals because of space limitations.

Another chemist, Jean-Claude Bradley (2009a, 2009b), put one of his students to work on another aspect of the replication. His communication of the negative findings in a blog (2009b) includes videos of the replication in progress and the resulting data. His separate video presentation about their failure to replicate (2009a) advocates for an innovation the Internet makes possible: open notebook science, committed to the display of all of the raw data previously tucked away in laboratory notebooks or hard drives. Bradley's video presentation (2009a) gives an excellent summary of the many ways scientists can exploit the Internet to open science to scrutiny by others. Full transparency to science in the making can become the norm in fields as different as neurobiology, astronomy, structural biology, and ethnography.

As a result of Docherty's and Bradley's efforts, efforts the Internet made possible, five months after its publication, *JACS* formally withdrew the article in question.[5] It might be argued that these dissenting scientists should not have published their refutation in a public blog; instead, they should have written to the paper's authors or to the journal in which they published. But this is to mistake their purpose; it was not to excoriate fellow scientists, but to open science to public scrutiny and critique. At the time of this writing, Docherty's posting on his blog led to no fewer than 151,445 views and provoked 211 comments on the experiments, the value of peer review, and the worth of Docherty's and Bradley's intervention. For example, a reader named "anniechem" praised Docherty, saying, "Fabulous post—chemists from all around the globe working together to see if this really could be true and reporting here. In less than 24 hours, some pretty damning evidence has been accumulated. Amazing!" In contrast, another

---

5. One of the most pernicious features of the Internet is the rapid and widespread transmission of misleading or erroneous information. But in the case of the research article, the Internet at least permits the rapid retraction of the information for anyone to see when accessing the article online, as exemplified by this case.

reader, "bigfish," was critical of Docherty's effort: "shouldn't there be some guideline on verifying the verification? I don't want to sound like Orwell's 1984 here. We all know who Tot. Syn. is and that he is qualified to do such thing, but let's say some guy shows up and tells us that La Clair's synthesis is irreproducible (or reproducible) on his blog. There's gotta be a more appropriate, for the lack of a better word, way of verifying someone else's work." In this case, anniechem seems to have the better of the argument. *JACS*'s action is impossible to conceive if Docherty and Bradley, rather than the authors of the original article, were doing sloppy science. In spirit if not in form, this exchange strongly resembles what Plato had in mind, a new form of intellectual exchange enhanced rather than undermined by writing and its accompanying images. In effect, a traditional self-regulatory mechanism, peer review, has been reinvented as a form of dialectical exchange, a form of intellectual collaboration in which relentless and repeated peer-to-peer interrogation is the self-regulatory mechanism.[6]

In the Internet collaborations we examine in the next section, while the mechanism of self-regulation varies, its purpose does not: it is always an embodiment of Robert Merton's (1979) norm of "organized skepticism," methodological doubt applied to all scientific work, one's own and that of others. In the first collaboration we look at, among scientists, each collaborator is responsible for the claims that are made and for the quality of the data and images on which those claims are based. The second and third collaborations we examine, between scientists and ordinary citizens, have a self-regulating mechanism of a different sort: a system of constraints embodied in computer software and the constant surveillance of professionals. In these examples, we see the start of what could be the revival of a new form of Platonic dialectic, a dialectic in which the scientific visualizations possible only on the Internet are integral.

COLLABORATION AND INTERNET VISUALIZATION

In *New Atlantis* (1627), Francis Bacon is remarkably prescient in envisioning the collaborative social structure most appropriate to the intellectual advance. The goal of science is trade *"not for Gold, Siluer, or Iewels; Nor for Silkes; Nor for Spices; Nor any other Commodity of Matter; But onely for* GODS *first Creature, which was* Light" (25). To achieve this goal, there is

---

6. See Harnad (1991) for one of the earliest discussions advocating open peer commentary by means of the Internet.

to be a division of labor between experimenters and theorists, and another between these two groups and "Nouices *and* Apprentices *that the Succession of the former Employed Men doe not faile; Besides, a great Number of* Seruants *and* Attendants, Men *and* Women" (40). Among natural philosophers there are to be "*diuerse* Meetings *and* Consults *of our wole* Number, *to consider of the former* Labours *and* Collections," as well as "Consultations, *which of the* Inuentions *and* Experiences, *which* [they] *haue discouered, shall be Published, and which not*" (45–46).

There is also a system of differential rewards based on achievement: "*For vpon euery* Inuention *of Valew, wee erect a* Statua *to the* Inuentour, *and giue him a Liberall and Honourable* Reward. *These* Statua's *are, some of* Brasse; *some of* Marble *and* Touchstone [a variety of quartz or jasper]; *some of* Cedar *and other speciall* Woods *guilt and adorned; some of* Iron; *some of* Siluer; *some of* Gold" (46–47).

In the seventeenth century, early scientific societies such as the Royal Society of London, Royal Academy of Sciences in Paris, and Lincean Academy in Rome all engaged in large collaborative projects; in the twenty-first century, the teamwork among scientists that Bacon envisioned has become the norm. Contemporary papers may have dozens, even hundreds of authors. In these collaborations, two forms of self-regulation are in place. One is traditional peer review. The other has joint responsibility as its base: since the reputation of each author is at stake, multiple authorship creates a web of mutual dependence in which men and women are held hostage to each other's credibility.

The Internet extends this web of mutual responsibility. In this section, we explore one of these extensions: the creation of a common evidential base. In "Impact of Scanner Hardware and Imaging Protocol on Image Quality and Compartment Volume Precision in the ADNI Cohort" (Kruggel et al. 2010), the central concern is finding the best scanner to trace patterns of brain deterioration in Alzheimer's patients. To discover this best scanner, these researchers analyzed 1,073 magnetic resonance images (MRIs) of 843 subjects, acquired on 90 scanners at more than 50 sites. The research team reached many conclusions, one being that among 17 scanner types tested, the optimal result was achieved with phased array (PA) head coils and field strength of 3.0 tesla (T).

Figure 8.1 compares the results of Philips Healthcare's Achieva PA 3.0 T and General Electric's Excite PA 1.5 T system. A careful examination of both image sets shows clearly that the Achieva is better than Excite at defining the boundary between white and gray matter, the relevant feature. We easily

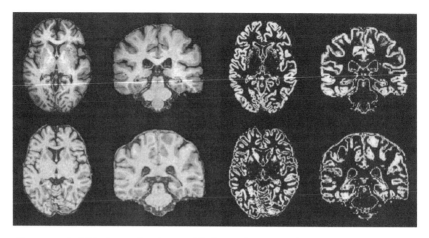

Figure 8.1. Axial (column 1) and coronal (column 2) sections of brain scan imaged by the Achieva system (top) and Excite system (bottom) (Kruggel et al. 2010, 2128). Columns 3 and 4 show the corresponding probability images of the gray matter class.

see that this is the case if in our mind's eye we superimpose the probability images on the right half of the figure onto the actual images on the left.

Although at first glance this article may seem typical of contemporary scientific collaborations, readers may have noted the uncommonly large number of examinations involved in its preparation: How many subjects? How many sites? Could one team implement a research design so broad in scope, so encompassing? In fact, the team recorded none of the MRIs, saw none of the subjects, and neither visited nor communicated with any of the sites. The final author in the article's byline accounts for these remarkable facts. This "author" is not a person, but a communal resource of data and images available through a web page open to all: the Alzheimer's Disease Neuroimaging Initiative (ADNI).[7] We have been examining a new kind of scientific article, one that uses as its resource base the results of a wide-ranging project involving the federal government, private industry, and university laboratories, under the direction of a single principal investigator, Dr. Michael Weiner, of the University of California–San Francisco. The range of this collaboration is wide indeed: pharmaceutical companies from Abbott and AstraZeneca to Lilly and Merck; medical device companies from General Electric to Medpace; the private philanthropic Foundation for the National Institutes of Health; the federal National Institute of Aging, a branch of the National Institutes of Health. The collaboration of universities

7. See http://www.adni-info.org/Scientists/LinksForDataCollection.aspx.

stretches from Mt. Sinai School of Medicine in New York to the University of California–Irvine; from the Mayo Clinic in Minnesota to the University of Alabama in Birmingham. In Canada, it extends from McGill University in Montreal to the University of British Columbia in Vancouver.

What motivated this extraordinary collaboration? It was clear to all participants that the challenge of Alzheimer's could not be met unless government, university, and pharmaceutical laboratories shared their data. There was a danger all could easily perceive: some scientists might publish before those who had actually gathered the data and images on which they relied, preempting them; there was also a danger that pharmaceutical companies might sacrifice both priority and profit. But all saw that, despite these dangers, their enlightened self-interest dictated a new form of collaboration. Their decision has been generally seen as correct. The project has been imitated in Europe, Japan, and Australia.

The ADNI is a collaboration from which amateurs are excluded. This was not always the case in science. In the nineteenth century, in geology, for example, talented and well-motivated amateurs actively participated in the scientific world. In *The Great Devonian Controversy*, Martin Rudwick indicates their place in the hierarchical structure of Victorian geology: elite geologists on top, amateurs on the bottom. Rudwick points out that geology was a special case. These amateurs

> were able to contribute far more substantially to geology than to most other sciences. Not only did they need no elaborate apparatus or equipment, their local situation gave them a positive advantage over the metropolitan or university geologists who might occasionally visit their home areas. Within those areas, amateur geologists had the *time* that was needed to find fossils by patient hammering in local quarries, cliffs, or cuttings ... and to build up a collection. In addition, as local gentry they soon became known to be willing to pay for any choice specimens that quarrymen and other laborers might come across from time to time. Their own reasons for amassing collections were often more aesthetic than strictly scientific, but provided they took care to record the localities where their specimens were found, they could have the added satisfaction of having their collections used and valued by some of the leading men of science of the country. (Rudwick 1985, 40; emphasis his; see also 17)

Although this description no longer characterizes geology, it captures the spirit of a contemporary collaboration between epidemiologists and their informants. In this collaboration, joint responsibility is out of the question. Instead, we have in place a self-regulating system of constraints embodied in computer software, a system supplemented by the constant surveillance of professionals. The paper "Information Technology and Global Surveillance of Cases of 2009 H1N1 Influenza" (2010), by John S. Brownstein et al., embodies this form of collaboration in three interactive images made possible by the Internet: a map (fig. 8.2), a line graph (fig. 8.3), and a scattergram (fig. 8.4). On the printed page, the map can represent the spread of H1N1 only by means of contrasting colors and geometrical shapes representing the stages of phase alert. A glance at figure 8.2 will demonstrate that this form of representation is far from transparent. In contrast, the interactive map we cannot reproduce makes such differentiations easy. By moving forward the time-scale cursor to a particular date, we obtain a snapshot of the state of the epidemic across the globe at that date; by moving the cursor forward at a steady pace, we watch unfold before our eyes the growth of the epidemic from start to finish.

Because its purpose is to reveal the epidemic's spread, the map has some of the qualities of a line graph: its circles and triangles index epidemic hot spots as they change over time. We can click on any of these geometric

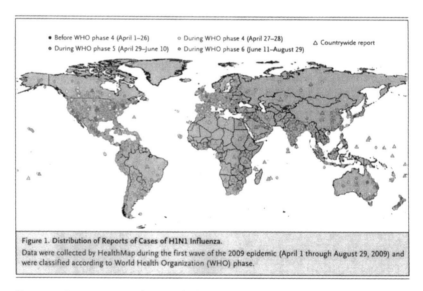

Figure 1. Distribution of Reports of Cases of H1N1 Influenza.
Data were collected by HealthMap during the first wave of the 2009 epidemic (April 1 through August 29, 2009) and were classified according to World Health Organization (WHO) phase.

Figure 8.2. Interactive map showing the location and spread of H1N1 (http://www.nejm .org/doi/full/10.1056/NEJMsr1002707). The World Health Organization has divided epidemics into 6 stages of alert, with stages 4 to 6 calling out for mitigation efforts. Reprinted with permission of the Massachusetts Medical Society.

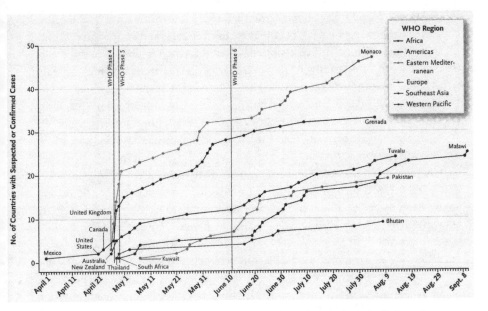

Figure 8.3. Interactive line graph showing the progress of H1N1 from April through September 2009 (http://www.nejm.org/doi/full/10.1056/NEJMsr1002707). Reprinted with permission of the Massachusetts Medical Society.

shapes; when we do, we obtain information concerning a source. For example, clicking a shape in the midst of the United States, we learn that the Associated Press has reported a story of March 11, 2010, with the headline, "Delaware Co. [Oklahoma] Girl Latest Swine Flu Death." Although the original map, published in the *New England Journal of Medicine*, had to remain intact so that it truthfully reflected the state of science at the time of publication, another version, at http://www.healthmap.org/nejm, was under no such constraint. Hosted by HealthMap, a website maintained by a large team led by the first author of the journal report and Clark Freifeld, a coauthor, it was kept up to date until July 2010, when H1N1 ceased to be of epidemiological interest. Still another version of this map, also hosted by HealthMap and found at http://healthmap.org, shows the location and spread of numerous epidemics, not just H1N1.[8] The diseases in this interactive map run the gamut from severe pneumonia to yellow fever.

Figure 8.3 is a line graph that indicates the growth of H1N1 from country to country. The time line begins with Mexico on the first of April 2009

8. Also see a video about the building of HealthMap at http://www.youtube.com/watch?v=vsp52VNVq5k.

and ends in September of that same year; trend lines, differentiated by color, trace the growth of each global region. In the line graph's interactive form, in contrast with its print version, detailed information concerning the epidemic's spread is readily available. Clicking on a particular country reveals a cluster that includes all reports on or around that date. For example, clicking on Chile reveals data for Argentina, Brazil, Columbia, Costa Rica, Uruguay, and Venezuela in the Americas, and Thailand in Southeast Asia: all of these countries reported confirmed cases on April 28. Moving the arrow below the figure across the time line on which it is superimposed, the reader can pinpoint the state of affairs at any confirmation date.

Figure 8.4 is the final graphic in the electronic version of the article by Brownstein et al., a scattergram not included in the printed article. It shows the relationship between national Gross Domestic Product (GDP) and time lag in reporting H1N1. In its interactive form this scattergram allows us to explore the relationship between the GDP per capita of each country and the time lag between the first formal report of a suspected case of H1N1 and a confirmed case. There is, as might be expected, a strong inverse correlation. Clicking on Austria, we find the gap is an acceptable single day; clicking on Zambia, the gap is an unacceptable 85 days. The authors explain:

> we explored the time difference between the issue dates of reports of suspected cases and confirmed cases of H1N1 influenza according to country. Overall, there was a median of 12 days (95% confidence interval [CI], 9 to 18) between reports of suspected and confirmed cases among the countries that reported these data. Lag times varied widely according to country. Large delays (up to 85 days) were probably due to a multitude of factors, including the reporting capacity of public health laboratories. We examined the association of this difference with the 2007 national gross domestic product (GDP) per capita (based on data from the United Nations Development Program), an approximate indicator of the strength of the public health infrastructure. . . . Countries with a high GDP per capita tended to have shorter lag times between issue dates of reports of suspected and confirmed cases (Pearson correlation coefficient, −0.4; 95% CI, −0.6 to −0.2), but there was wide variation in lag times among less affluent nations. (Brownstein et al. 2010, 1732)

All three of these interactive graphics differ from their printed versions in that the reader has many more paths for extracting meaning from them: they are the visual equivalent of hypertext. Still, the Internet and print

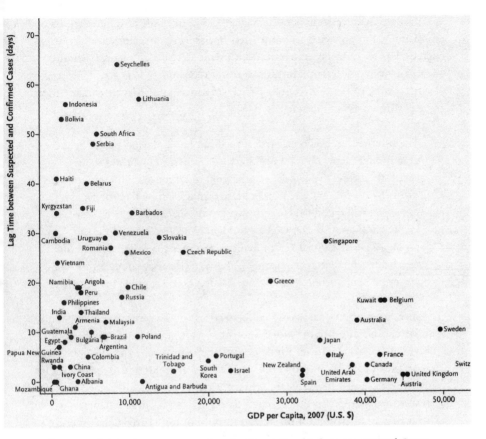

Figure 8.4. Interactive scattergram showing the relationship between national Gross Domestic Product (GDP) and time lag in reporting H1N1 (http://www.nejm.org/doi/full /10.1056/NEJMsr1002707). Reprinted with permission of the Massachusetts Medical Society.

versions share a common purpose: a significant pattern among the acquired data lies hidden until the graphic realizes it visually. To bestow meaning on these patterns of data, however, only words will do, reports from all over the world. Where do these originate? Health care workers are one source, but they are not the only, nor necessarily the most important. Local media and the general public also participate. For example, when we click on "Cameroon," we find a message in French from "smerkaru": there have been "23 morts dans une épidémie de choléra dans l'extrême-nord." Smekaru's report of these deaths in the extreme north of his country, based on a newspaper story, has been read 114 times. His is not a special case: anyone can add a comment via e-mail, voice mail, text, or smart phone submission. A news story can be shared, or an eyewitness account. Readers can passively participate in information gathering; 114 did. But smekaru chose to be an informant. In supplying this information, an ordinary citizen without expertise has become a member of a worldwide network of medical scientists, health workers, government officials, and reporters whose goal is to reveal the presence of epidemics in a timely manner, and thus to control their spread. Brownstein summarizes:

> The Internet has become a critical medium for clinicians, public health practitioners, and laypeople seeking health information. Data about diseases and outbreaks are disseminated not only through online announcements by government agencies but also through informal channels, ranging from press reports to blogs to chat rooms to analyses of web searches (see Digital Resources for Disease Detection). Collectively, these sources provide a view of global health that is fundamentally different from that yielded by the disease reporting of the traditional public health infrastructure. (Brownstein, Freifeld, and Madoff 2009, 2153)

In contemporary scientific collaborations on the Internet, the general public can be more than informants; they can be active participants (see also Nielson 2011). Our example concerns the unraveling of the structures of complex proteins. In molecular biology, as in architecture, form follows function: to cure a disease, a pharmaceutical must be synthesized that fits exactly into a culprit protein, blocking its action. At first, scientists at the University of Washington interested in protein structure sought merely to harness the collective power of home computers, creating among citizen volunteers a distributed super-computer with a capacity ample enough to meet the challenge of complex structures. To accomplish this task, they created Rosetta@home, a software program that appears on home computers as a screen saver, while behind the scenes it crunches data to solve the problem of "how a linear chain of amino acids curls up into a three-dimensional shape that minimizes the internal stresses and strains—presumably the protein's natural shape" (Hand 2010, 685). But the program solves these problems in the only way software can. Loaded with existing possible configurations, it plods methodically from one to another, trying over time to reach the nirvana of the protein's "native state." In this state, the water-loving components will be on the outside of the molecule, the water-hating ones on the inside, and the energy level will be as close to zero as possible. Under such circumstances, the molecular representation is as stable as the molecule itself: the native state has been reached.

At this point in its development, the project was interesting but not innovative; for example, Einstein@Home had already discovered a new pulsar by means of a similar volunteer computer network. But then came a moment of creative insight, not on the part of the Rosetta@home scientists themselves, but on the part of some of their citizen volunteers. Staring at their screen savers, it occurred to some that they might be able to outperform the plodding software. This collective insight energized the scientists involved. They began to wonder whether they could harness not only the computers of their volunteers, but their visual intelligence as well. So motivated, the scientists invented Foldit, a spatially oriented computer game on the lines of *Tetris*. For the most part, its players were not scientists; they simply enjoyed the challenge of the game. In a video produced by *Nature*,[9] a top English gamer named Susanne, a woman with no background in biochemistry, says, "when I go home [from my job as a rehab administrator], I am a different person," exercising "abilities I didn't know I had." David Baker, a lead scientist on the project, says he thinks people, when asked at the office what they did last night, might prefer to say not that they had played *Halo*, a science fiction game, but that they helped find a vaccine for HIV. In this, as in the Brownstein et al. collaboration, joint responsibility is out of the question. Instead, we see in place a self-regulating system of constraints embodied in computer software, a system supplemented by the constant surveillance of professionals.

What these gamers see in screen representations like figures 8.5 and 8.6 is a molecular "cartoon," a symbolic representation designed to show the spatial relationships among the atoms of a problem molecule. Different visualizations of the molecule—ribbon and ball-and-stick cartoons, for example—reveal different aspects of its structure. No one can actually see the complex molecule as it "really" is. Nonetheless, through computer simulations gamers can see visual representations of these structures that scientists routinely create. In their spare time, they face a screen like figure 8.5. In this figure, numbered arrows point to the following screen features (Cooper et al. 2010, 757):

- Gamers see the characteristics of the molecule. They see atoms that are too close (labeled as "1"); hydrogen bonds (2); a water-hating side chain with a yellow blob to show that it is exposed when it ought to be hidden inside the structure (3); a water-loving side chain (4); a segment of the backbone, colored red because it has unacceptably high energy (5).

9. See http://www.nature.com/nature/videoarchive/foldit/index.html.

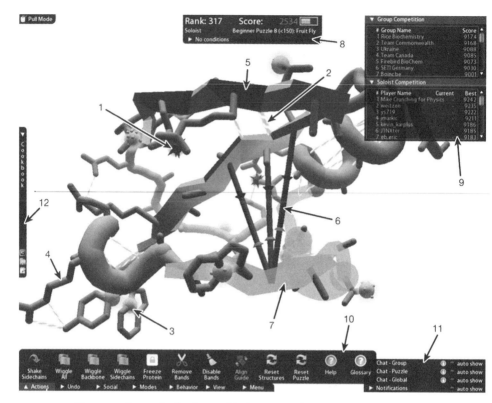

Figure 8.5. Screen display from Foldit (Cooper et al. 2010, 757). Reproduced by permission of Nature Publishing Group.

- They see ways of manipulating the molecule: by means of "rubber bands" (6) and by freezing (7), which prevents degrees of freedom from changing.
- They see the user interface: players' status and score (8 and 9); tool bars for accessing tools and options (10); chat (11); and a "cookbook" for making new automated tools or "recipes" (12).

Although it is too early to know whether Foldit will lead to significant scientific discoveries, a leading scientific journal has already taken notice of the project's potential. In an article in *Nature*, Cooper et al. (2010) point to the project's many achievements. For example, figure 8.6 shows the results of a molecular puzzle in which players have outdone the computer program. In figure 8.6*a*, a black dot represents the initial state of the molecular puzzle (placed at about 4.3 angstroms). The Rosetta program's performance is in yellow (lighter shade); Foldit's, in green (darker shade). We can see at a glance that the winning Foldit player (lowest scoring Rosetta energy) did better: in fact, its prediction was only 1.4 angstroms away from the native state at zero, whereas the Rosetta program best was 4.5 angstroms away. In figure 8.6b, we see the result in the form of a molecular cartoon. The puzzle is in red, the native state in blue. In the color version, it is easy to see that the green solution of Foldit is closer to the molecule's native state than the yellow solution of Rosetta. There were ten attempts at puzzle solving; the players did better than Rosetta in five, whereas Rosetta's predictions were more accurate in two. There were three ties.

In the researchers' view, gamers succeeded because of their ability to persist in counterintuitive maneuvers. In a particular instance,

> Rosetta's rebuild and refine protocol . . . was unable to get within 2A °
> [angstroms] of the native structure . . . This example highlights a key

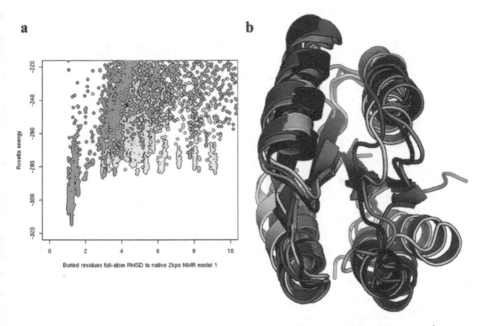

Figure 8.6. Results of Foldit players and computer in molecular modeling (Cooper et al. 2010, Supplementary Information, 13). The y-axis is the Rosetta energy; the x-axis, the root mean square deviation from the "native state." Reproduced by permission of Nature Publishing Group.

difference between humans and computers . . . [S]olving the strand-swap problem required substantially unraveling the structure, with a corresponding unfavourable increase in energy. Players persisted with this reconfiguration despite the energy increase because they correctly recognized that the swap could ultimately lead to lower energies. In contrast, although the Rosetta rebuild and refine protocol did sample some partially swapped conformations, these were not retained in subsequent generations owing to their relatively high energies, resulting in the top-scoring Rosetta prediction being further from the native than the starting structure. (Cooper et al. 2010, 757–59)

The Foldit video ends with one of the gamers predicting that their strategies will eventually be incorporated into the software, rendering gamers redundant. This is represented visually by a panel in which the gamers disappear, one by one, leaving a blank screen. But this prediction of intellectual suicide may be overly pessimistic; it is not clear at this point that computer programs can completely reproduce the human brain's problem-solving ability.

VIRTUAL WITNESSING AND INTERNET VISUALIZATION

Like collaboration, virtual witnessing has its roots in the ideology of early modern science. Those who had access to the laboratory or lecture room where an experiment was conducted could see for themselves; those who did not could experience these same events vicariously by means of a verbal tsunami, designed to re-create these events on the printed page. In the words of Robert Boyle, the inventor of virtual witnessing, "these narratives [are to be] as standing records in our new pneumatics [study of materials in near-vacuum], and [readers] need not reiterate themselves an experiment to have as a distinct idea of it as may suffice them to ground their reflections and speculations upon" (quoted in Shapin and Schaffer 1985, 62). Whether this is the case, or whether we have in Boyle's writing the faithful transcription of the running speech of a purblind and loquacious gentleman, transcribed by a diligent amanuensis, we probably cannot know; in any case, this difference can be consigned to the arena of historical debate. What is not in doubt is that the descriptions and depictions of scientists then and now exhibit far more detail than those of the novelist or the ordinary citizen, and that they do so in the interest of epistemic candor. It seems incontestable that the Internet visualization has permitted the virtual witnessing Boyle invented to take a gigantic leap forward in the direction of increased credibility. In our examples, scientist writers display their evidence to their readers in ways not possible in print: moving pictures of radial glia, computer simulations of biological molecules and star explosions, and audio recordings of the songs of mice.

An article by Hansen et al., "Neurogenic Radial Glia in the Outer Subventricular Zone of Human Neocortex" (2010), exemplifies this epistemic advance. It sets out to discover the aspect of the developmental process by means of which the evolutionary leap in neuronal capacity arose in the human brain. Figure 8.7 makes clear the stark neuronal contrast with the mouse: in humans, the expansion of the subventricular zone is so great that it can be represented only by a jagged gutter (immediately below the cortical plate, CP), a conventional representation of otherwise unrepresentable

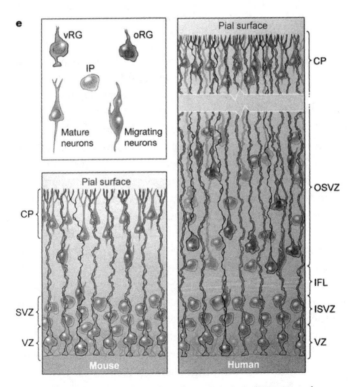

Figure 8.7. Visualization of mouse vs. human neurogenesis (Hansen et al. 2010, 559). vRG = ventricular epithelium cells; oRG = radial glia-like precursors of the outer region subventricular zone; IP = intermediate progenitor; CP = cortical plate; IFL = inner fiber layer; OSVZ = outer region subventricular zone; SVZ= subventricular zone; VZ = ventricular zone; ISVZ = inner subventricular zone. Reproduced by permission of Nature Publishing Group.

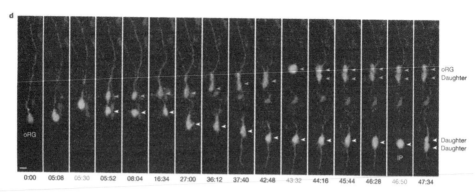

Figure 8.8. Animation of biological process: three-stage proliferation of radial glia-like precursors in the outer region subventricular zone (oRG) (Hansen et al. 2010, 556). IP = intermediate precursor. Scale bar on left represents 10 micrometers. Animated version available in article's online supporting information. Reproduced by permission of Nature Publishing Group.

space. The human subventricular zone differs importantly from an analogous one in the mouse in that it contains numerous radial glia-like cells that contribute in a significant way to the creation of neurons ("oRG" in the image displayed in upper left). Employing real-time imaging and clonal analysis of fetal tissue from voluntarily terminated pregnancies, Hansen et al. find that the radial glia-like precursors of the outer region subventricular zone depicted in figure 8.7 can self-generate. In figure 8.8, this proliferation is exhibited in two stages, a rapid generation that may account in part for the complexity of the human cortex. In frame 05:52 [minutes], we see a first proliferation of an oRG cell into a daughter oRG cell. In frame 47:34, the initial oRG (intermediate progenitor, IP) has divided into two daughter cells. We thus begin with oRG on the far left; at the end of the process, on the far right, we have oRG next to a daughter cell, underneath of which are two further daughter cells. We cannot, however, actually experience this process of self-renewal unless we animate these consecutive stills, in our mind's eye filling in the blank spaces between them. Instead of engaging in this nearly futile imaginative effort, we can turn to the authors' supplemental materials, where four short movies allow us to see the process in action, in other words to see exactly what the researchers saw.

New-style virtual witnessing can also animate objects. Tobias Karlberg and his research team published a communicatively conventional scientific article with thumbnail images in *PLoS One*, "Crystal Structure of the ATPase Domain of the Human AAA+ Protein Paraplegin/SPG7" (2010). By visualizing and analyzing the structure of the paraplegin gene, they had dis-

covered the source of the mutations that cause spastic paraplegias, a relatively rare inherited disease marked by progressive stiffness and contraction
in the lower limbs. To best display the 3-D rotatable structure they had
derived by means of computerized model building based on their experimental findings, Karlberg and his coworkers also published a state-of-the art
version of the same article, a version possible only on the Internet.[10] In this
version of the article, the screen is divided in two halves, with a scroll bar in
between. On the left, we see the same text and images that appeared in *PLoS
One*; on the right, we see an interactive 3-D molecular map of paraplegin.
As we scroll down the left side, the article text and image progress just as
with any other Internet article, but the molecular map on the right side remains stationary. As we scroll down, however, we can also click on links to
noun phrases in the text such as "crystal structure of the AAA-domain of
human paraplegin bound to ADP":[11] an image similar to figure 8.9 appears
on the right half, a translation of the noun phase into an interactive 3-D
image. In this image the tinker-toy structure in the middle is the ADP, and
the ribbon structure hovering around it is the AAA domain. The main components of the AAA domain include two α-helical bundles (α1–α4 on the
left side and α5–α8 on the right side) and the "five-stranded parallel β-sheet"
(β1–β5). Such phrases linked to visualization of the science appear throughout this Internet article. If we so desire, we can also move our cursor to the
image and rotate the structure, zoom in or out, change the background from
black to white, or view the structure in sixteen different biological settings.
From their visual study, Karlberg et al. concluded that the key contributor
to disease mutations is the part labeled Δ485–487 (upper right).

New-style virtual witnessing of course extends beyond biology. In
"Three-Dimensional Simulations of Mixing Instabilities in Supernova Explosions," the research team of Hammer, Janka, and Müller (2010) used 3-D
computer simulation to re-create, for the first time, the explosion of a supernova "from the first hundreds of milliseconds to a time 3 hr later when
the shock has broken out of a blue progenitor star" (1372). Visualization of
the 3-D simulation in print, which calls for a spatiotemporal representation, is restricted to a progression of static images at different times, as in

---

10. See http://www.plosone.org/enhanced/pone.0006975; download of plugin required.
Also, for a much better sense of the new form of verbal-visual interaction the Internet permits,
see the video created for this PLoS article at http://www.youtube.com/watch?v=ZUi3yltVTbU.

11. AAA stands for "ATPase associated with diverse cellular activities"; this protein is
central to many biological processes, such as protein degradation, DNA replication, and gene
expression regulation. ADP stands for adenosine diphosphate, an important molecule involved
in energy transfer in cells.

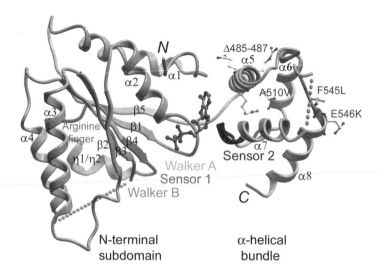

Figure 8.9. Animation of biochemical object: Structure of paraplegin gene (Karlberg et al. 2010). Animated version available at http://www.plosone.org/journals /journalNamePlaceholder/webapp/enhanced/pone.0006975.

figure 8.10. The upper two images model the supernova surface from two viewing perspectives after about six minutes postexplosion, and the bottom two panels, the explosion after three hours postexplosion. But these are isolated frames from a feature-length film; an Internet video allows us to view the entire three-hour supernova explosion in a time-lapse simulation that runs for about half a minute.[12] The author's visual study revealed important differences between their 3-D model and earlier 2-D versions.

Internet virtual witnessing can also include an auditory dimension. In "Ultrasonic Songs of Male Mice," Timothy Holy and Zhongsheng Guo (2005) explore whether the sounds mice make, sounds beyond the human hearing range, really are songs. In the strictest sense, songs are patterns that cohere to form organic wholes persisting over time. Bird vocalizations exemplify song in this strictest sense; Holy and Guo conclude that mice vocalizations almost do. These "contain multiple syllable types, and these syllables are uttered in regular, repeated temporal sequences" (2182); moreover, "individual males also have characteristic temporal structure to their songs" (2183). The researchers infer from this that "these songs satisfy [the] *sensu stricto* definition of song . . . , as well as many aspects of [the] *sensu strictissimo*" (2183). In the printed paper, figure 8.11 makes their point visu-

12. See video at http://www.youtube.com/watch?v=8YFov4qOByA.

ally; like a musical score, this image shows striking parallels among three mouse song sequences: each phrase begins with two or three "SS" syllables followed by six to eight "du" syllables. However impressive these parallels are, they cannot match in vividness our Internet experience of "hearing" the mice, a series of four recordings followed by a recording of the song of the juvenile swamp sparrow. At first, we hear nothing remarkable. The mice sound like mice, their squeaks forming the patterns that Holy and Guo have revealed. It is only when we hear the birdsong that the auditory revelation occurs: the parallel between it and mouse song is compelling. Figure 8.11 and the experience of the recordings dovetail, leading to the

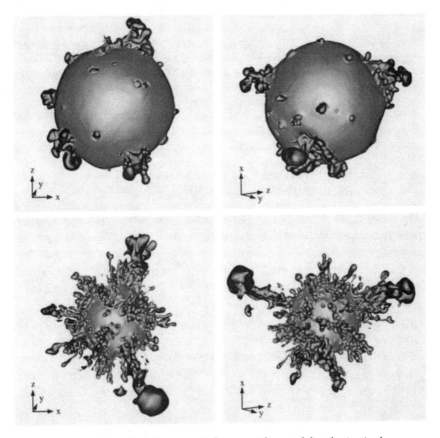

Figure 8.10. Animation of astronomical process: Elemental distribution (carbon, oxygen, and nickel) during supernova explosion (Hammer, Janka, and Müller 2010). Upper two images, at 350 seconds; lower two images, at 9000 seconds. Animated version available at http://www.youtube.com/watch?v=8YFov4qOByA. Reproduced by permission of the AAS.

Figure 8.11. Visualization of sound: phrase repeated three times without interruption in the mouse song (Holy and Guo 2005, 2181). The three repeats are shown one above the other, aligned on the start time for the phrase. Audio recording supplements visual.

same conclusion. In this case, perception and cognition fuse to form the firmest of evidential bases, a firmness only the Internet makes possible.

DEMONSTRATION AND ONLINE VIDEOS: INFORMING AND EDUCATING OTHERS

The Internet also hosts short videos that keep specialists and nonspecialists informed about the latest discoveries and instruct specialists in the performance of highly technical procedures. Such videos are routinely available in the elite scientific journals as a complement to their highly specialized research articles. In no discipline is this more evident than in medicine, which differs in social structure from other scientific communities. In medicine, a small core of researchers serves the new knowledge needs of a much larger number of practitioners.

The task of keeping the latter abreast of the latest research had become increasingly difficult in our age of ever-increasing specialization. To meet this challenge, most issues of the *Journal of the American Medical Association* contain an article of wide interest to the medical community accompanied by a two-minute informational video. An example is Lindenauer et al.'s "Association of Corticosteroid Dose and Route of Administration with Risk of Treatment Failure in Acute Exacerbation of Chronic Obstructive Pulmonary Disease [COPD]" (2010). The title of the accompanying video incorporates the article's claim: "equal or better outcomes [are realized] for hospitalized COPD patients when given low versus high dose steroid therapy."

While online videos may be relatively new, their structure is not. These are organized along the lines of a familiar genre, the news story. This particular video opens with a shot of a patient, Francis Welch, who suffers from COPD, or chronic obstructive pulmonary disease. The retired dentist is shown seated in an examining room, speaking to his doctor. He is hooked up to an oxygen supply designed to help overcome his breathing problems. His story of a lifelong addiction to cigarettes humanizes COPD, a disease that forces Dr. Welch to "think about breathing sixty minutes every hour." After some brief changes of scene—we see breathing tests, exercise programs, a hospital patient—the study's chief author, Dr. Lindenauer, is introduced. He reiterates the title's claim, a conclusion based on a study that included more than 400 hospitals and nearly 80,000 patients. A lower dosage of steroids, the doctor asserts, avoids their unpleasant side effects without negating their benefits. The presentation closes by returning to Dr. Welch, who warns about the dangers of smoking. The video has the elements typical of narratives. The characters are the patients and the researcher; the time is the present; the place is the examining room, the exercise room, and the hospital; the problem is that COPD symptoms have been routinely alleviated by intravenous injection of high doses of steroids, which have unpleasant side effects; the resolution is a lower dose of steroids administered orally, a treatment that is just as effective; the moral is, Do not smoke.

JAMA videos are policy-neutral: they avoid commenting on the articles from which they are derived and never voice any criticisms of the medical community or of medical policy that the articles may state or imply. This is not pusillanimity; it is a recognition that these videos speak with an institutional voice, that of the American Medical Association. The avoidance of institutional endorsement for critical and dissenting views, however, does not mean that criticism or dissent is avoided; it means only that it is the job of individual contributors, not the Association. As a consequence, there is a division of communicative labor between the video and the texts accompanying it. The former vividly depicts the central theme of the article in human terms, the job of the standard news story; the latter are more analytical and critical, jobs newspapers assign to investigative journalists and the op-ed page. In an editorial accompanying Lindenauer et al., Krishnan and Mularski (2010) point out that high steroid dosage is an established practice hard to dislodge, regardless of the evidence; in any case, they feel, more evidence is needed to establish Lindenauer's procedure as legitimate clinical practice. So in this case we have a multitiered communication: a traditional scientific article and its web-based version, a video dramatizing the central point of the article, and a critical commentary.

The medical community is now using the Internet not only to inform, but to educate. In regularly presenting its diagnostic case studies, *The New England Journal of Medicine* continues a tradition in medical education, one that has evolved over time from a practice Hippocrates long ago inaugurated (Reiser 1991). "Case 17–2010: A 29-Year-Old Woman with Flexion of the Left Hand and Foot and Difficulty Speaking," by Tarsy et al. (2010), adds to the traditional case study an etiology grounded in genetic defect. In this study, words, video, and still images work together to re-create for an audience of readers the experience of a live audience participating in a grand rounds. In the video, the patient exhibits her symptoms. We peer into her oral cavity to see her malfunctioning vocal cords. In figure 8.12, a frame from the video, we also see her awkward gait. What the video reveals, the article analyzes. It presents a stylized version of the argumentative exchange among the senior physicians, a dialectical encounter that leads, finally, to a diagnosis of "rapid-onset dystonia-parkinsonism due to a mutation of the ATP1A3 gene" (2217). Figure 8.13 presents readers with a diagram depicting the genetic malfunction at the root of her disease. The mutated amino acid residue appears as a conspicuous yellow blob, which impairs the pumping of sodium ions (Na^+) out of the cell and potassium ions (K^+) into it. It is this malfunction that is the cause of the awkward gait revealed in figure 8.12. Readers see the patient and her presenting symptoms in a video; read an exchange among physicians diagnosing the symptoms, naming the disease and its genetic cause; and see a color illustration of the genetic cause.

Figure 8.12. Animation of medical condition: Awkward gait of a patient with dystonia-parkensonism (Tarsy, Sweadner, and Song 2010). Video accessed at http://www.nejm.org /doi/full/10.1056/NEJMcpc1002112.

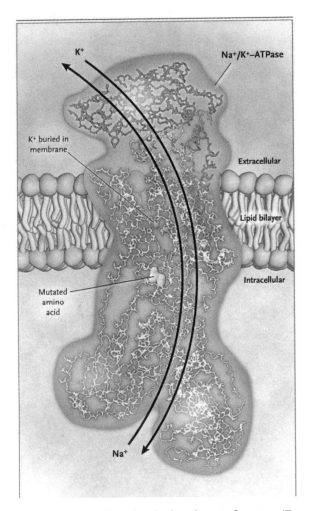

Figure 8.13. Visualization of cause behind medical condition in figure 8.12 (Tarsy, Sweadner, and Song 2010, 2218). Reprinted with permission of Massachusetts Medical Society.

The New England Journal of Medicine also provides instruction concerning preferred medical procedures. An example is Shaikh et al.'s "Diagnosing Otitis Media: Otoscopy and Cerumen [earwax] Removal" (2010). This eight-minute video, a form of user manual, contains two categories of information: declarative and procedural. By means of the former, the instruments needed are described; by means of the latter, their use is illustrated (Mackiewicz 2005). "Equipment" is a sequence within the video devoted to the instruments needed for cerumen removal and pneumatic otoscopy. In

an early frame, one we do not depict, we see a list of the instruments needed
(otoscope, speculums of various sizes, blunt ear curette, and triangular applicator) accompanied by their pictures. As the otoscope is being discussed,
its picture and the word "otoscope" are simultaneously bolded to reinforce
recognition. The voice-over adds to this reinforcement: "For pneumatic
otoscopy," it says, "you will need an otoscope with a diagnostic head that
has an attached rubber bulb and a movable lens." In addition, the voice-over
sends a message no picture can: for cerumen removal, physicians should
prefer an otoscope with a surgical head. This declarative visual information
is a necessary prelude to the procedural information that follows.

To illustrate the depiction of procedural information, we analyze four
frames on cerumen removal, each of which facilitates the learning necessary for "appropriate actions." The first, reproduced in figure 8.14, focuses
on a modification of a triangular applicator, just listed in "Equipment."
Carefully twisting cotton around the applicator's tip, the physician creates
a tool that can be used to sweep up residue, once the bulk of the cerumen
has been removed. The angle chosen for this depiction is just above the
physician's hands as she wraps the cotton around the applicator; not coin-
cidentally, this is the same angle from which any physician learning this
procedure must view the applicator.

Figure 8.15 also facilitates learning by showing exactly what the physician sees as she hooks the loop of the blunt curette behind the cerumen mass.
Figure 8.16 shows us that it is only from a view a physician cannot possibly

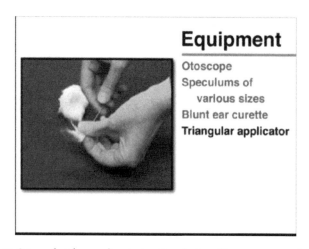

Figure 8.14. Screen shot from online instructional video: How to create an instrument
for residual cerumen removal by wrapping cotton around a triangular applicator (Shaikh
et al. 2010). Video accessed at http://www.nejm.org/doi/full/10.1056/NEJMvcm0904397.

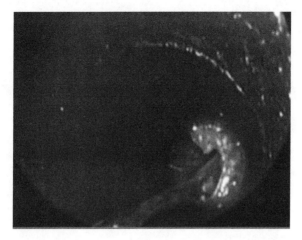

Figure 8.15. Screen shot from online instructional video: View through the otoscope showing a blunt curette removing cerumen from ear canal (Shaikh et al. 2010). Video accessed at http://www.nejm.org/doi/full/10.1056/NEJMvcm0904397.

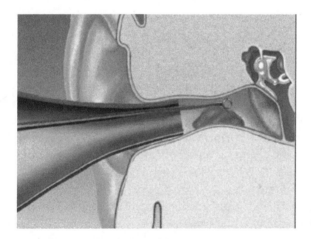

Figure 8.16. Screen shot from online instructional video: How to use the loop of a blunt curette to hook cerumen from behind (Shaikh et al. 2010). Video accessed at http://www .nejm.org/doi/full/10.1056/NEJMvcm0904397.

see that the procedure photographed in figure 8.15 can be sufficiently clarified. This animation sequence reveals that the loop of the curette must be hooked *behind* the cerumen mass and dragged forward out of the ear. In so doing, care must be taken to stay well clear of the tympanic membrane and the ear canal. This depicted caution is reinforced in a voice-over: "Try to pass the loop of the curette *behind* the piece of cerumen, being careful not

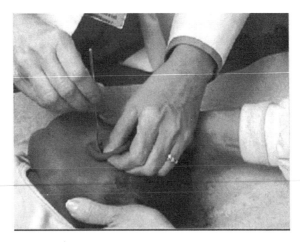

Figure 8.17. Screen shot from online instructional video: How to remove residual cerumen with a triangular applicator wrapped in cotton (Shaikh et al. 2010). Video accessed at http://www.nejm.org/doi/full/10.1056/NEJMvcm0904397.

to traumatize the walls of the external auditory canal or the tympanic membrane" (spoken emphasis theirs). Figure 8.17, a frame from the conclusion of the cerumen removal procedure, allows viewers to see what a physician in training sees as the procedure is executed by another physician. This is also an angle directly facilitative of learning. Voice-over supplements the view's visual information; it tells the physician in training to insert the cotton-wrapped applicator about five centimeters into the ear canal.

Accompanying the video is a link to a verbal description with still pictures, and a long editorial on the importance of accurate diagnosis for infants and children. Whether the website visitor is a verbal or visual learner, appropriate information is online to help.

In sum, we have analyzed three web-based genres designed to inform and instruct—a video news story, a video presentation of a diagnostic conference, and an instructional video teaching a procedure. Each focuses on a single topic; each exhibits close coordination between the verbal and the visual; each displays "effects that could not be achieved in print" (Hayles 2008, 59).

THE SCIENTIFIC JOURNAL AND INTERNET VISUALIZATION

In a quantitative study of the current scientific literature, *The Scientific Article in the Age of Digitization*, Owen (2005) analyzed eighty-six peer-reviewed scientific journals that had initiated publication of digital research articles between 1987 and 2001. His principal conclusion was that "contrary to pretentious claims and expectations about the impact of digitization on scientific communication (5) . . . , the journal article as a communicative form for reporting on research and disseminating scientific knowledge does not seem to have been transformed by . . . [the Internet]: *it remains a digital copy of the printed form*" (11; our emphasis). Owen views the current situation as preserving and extending "existing functions and values rather than as an innovation that radically transforms a communicative practice that has evolved over the centuries" (55). In other words, gradual evolutionary, not revolutionary change, has occurred in the communicative aspects of the research article in the typical digital journal.

We have no reason to doubt the validity of Owen's conclusion regarding his sample at the time of his book's publication. However, not only is his sample more than a decade old, but it also included such journals as the *Brazilian Electronic Journal of Economics, Internet Journal of Chemistry*, and *Journal of Cotton Science*, and not the most highly cited scientific journals producing printed and electronic issues, nor the highly successful open-access journals founded by the Public Library of Science (PLoS). Our contention is that the future is largely in the elite scientific journals such as *Science, Nature, Cell, PLoS Biology, Journal of the American Medical Association*, and *New England Journal of Medicine*.[13] We maintain the Internet is dramatically changing the visual representation of the article as a whole; what was the first page of a print article has become a gateway to its sections, to supplementary materials, to related articles, and to information about its impact (see, for example, fig. 8.18).

In the Internet-only article, figures and tables can be positioned immediately after first mention, not where the exigencies of the printed page require. For each figure, readers have the option of either ignoring it altogether, clicking on a link to view it in a much more easily scrutinized size, copying the figure for use in another electronic document, or viewing and copying the figure as a PowerPoint slide. Some elite journals also prominently display all the figures within an article in a separate segment on the screen at the article's top or side. For example, *Nature* now displays all figures in a horizontal band after the abstract and before the introduction.

---

13. Elsevier has a website posting what it calls prototypes for "the article of the future" (actually, mostly features now available in the avant-garde scientific journals) at http://www .articleofthefuture.com.

Figure 8.18. Screen shot of digital article in 2010 *PLoS Biology* (Aron et al. 2010, http://
www.plosbiology.org/article/info%3Adoi%2F10.1371%2Fjournal.pbio.1000349).

With this arrangement, before jumping into a sea of specialized terms and
data, the viewer can read the abstract, scroll through all figures, and quickly
make a judgment as to the article's value and relevance.

The visual display of reference information has also radically changed
the intertextuality of the scientific article. Subject to the ever-loosening restrictions on copyright by publishers of scientific books and journals, references can be "live"; views of cited articles can be only a click away. These
references might include links not only to the citations in the reference
list, but also to published works in the same journal by the same authors,

uncited articles on the same topic by other authors in the same journal, articles related to key words chosen by the authors, information about manufacturers of equipment and materials used in experiments, relevant databases such as the Protein Data Bank, articles that cite the subject article after publication, written comments by readers and the authors' responses, and even blogs on the subject matter in question. In the case of "newsworthy" articles, general reader video presentations and podcasts, including interviews with the scientists involved, can be accessed in supplementary materials. In these, we can also see much that has been excluded from public view by the exigencies of print: for example, supplemental tables of data, supplemental methods without page restrictions, and film clips of field and laboratory events.

The *intra*textuality of the digital scientific article has also altered its visual landscape. As pointed out by Gross, Harmon, and Reidy (2002), one of the more remarkable characteristics of the twentieth-century scientific article is its elaborate finding system of section headings, graphic legends, numbered citations, and numbered equations, a system that evolved gradually over time. This master finding system allows readers to navigate more easily among the diverse components of the article; they can extract desired bits of theory, methods, results, and conclusions without having to read the text from front to back. The finding system acts like a map, allowing readers easily to direct their attention to select components within the argument. In the digital scientific article, this finding system has been augmented by digital links from section headings to actual sections, from figures or tables called out in running text to actual figures or tables, from citation numbers in the text to actual references. Moreover, at each internal heading, a dropdown window can be opened with links to all sections. The element of greatest interest to any given reader is just a click away.

To widen the potential audience, some journals have experimented with the visual display of summary information. For example, the PLoS journals display not only the usual heading abstract, but also an "author summary" of about the same length whose stated purpose is "to make findings accessible to an audience of both scientists and non-scientists . . . Ideally aimed to a level of understanding of an undergraduate student."[14] Some journals also are taking advantage of the ease of displaying images and text to catch the readers' eyes: the journal contents page shows the usual lists of titles and authors, but also a one- or two-sentence abstract of the abstract and the

14. PLoS Biology Guidelines for Authors, http://www.plosbiology.org/static/guidelines .action.

key image, usually in color.[15] Other journals' contents pages offer "multimedia centers" composed of scientific images displayed for the reader's aesthetic pleasure or edification or both, podcasts highlighting the contents of
individual issues or other newsworthy matters, instructional webinars and
visuals, slideshows on a variety of scientific topics, and short videos produced by the journal or by authors of articles.

Worth mentioning also is the *Journal of Visualized Experiments* (*JOVE*),
which could one day revolutionize the methods section in scientific articles. Moshe Pritsker founded it in 2006 to resolve a key problem in the
life sciences—the extreme difficulty in reproducing experiments based only
upon written information. *JOVE* overcomes this difficulty by taking "advantage of video technology to transmit the multiple facets and intricacies
of life science research." Each article begins with a professionally produced
video (5 to 15 minutes) demonstrating some experimental method in the
life sciences, such as the quantification of DNA in small concentrations.
Video-recorded methods allow viewers to see exactly what the authors did
in the laboratory, and also, significantly, what they avoided doing. By these
means, viewers can more readily tease out many of the assumptions seldom spelled out in printed methods sections. But *JOVE* does not subsist by
video alone: following each video there appears a detailed abstract describ-
ing the method and its application; a step-by-step protocol, often including
additional visuals; and representative results combined with a discussion of
their significance.

In the past, scientists wanting to replicate a difficult experiment had
to spend weeks or months learning and perfecting a new technique, often
having to visit the authors' laboratory for a demonstration. The goal of this
journal is no less than greatly reducing this major inefficiency in scientific practice by creating a vast web-based repository of "visualized experiments" that feature video streams of scientists at work in the lab. These are
intermixed with processes captured by video microscopy and microscopic
structures rendered by 2-D and 3-D animation.

CONCLUSION

At the beginning of this book, we presented a model of the diverse ways
the verbal and the visual interact to create and communicate science: in

15. Though color images have appeared since the earlier days of scientific publishing, because of cost in production and printing, they were rarely used through the twentieth century; it
is only with the introduction of the electronic journal that they have now become routine.

essence, viewers move from perception of an image to identification of its components, to interpretation of those components, and, finally, to the integration of the visual into scientific arguments and narratives. Each step must precede the one following—we cannot possibly identify what we have not first perceived, nor interpret what we have not first identified. Of course we can backtrack at any time: we can reidentify what we have already interpreted and, as a result, reinterpret it. In the remainder of the book, we applied our model to real-world examples of scientific visuals appearing in different communicative contexts: journal articles and books, as well as PowerPoint presentations, always emphasizing the vast majority of cases in which scientific meaning is the product of verbal-visual interaction.

In this last chapter, we argued that the Internet is radically transforming scientific visualization and, consequently, the speed and complexity of verbal-visual interaction. First, the Internet is making possible a new form of Socratic dialogue between researchers and their readers. Unlike the traditional Socratic dialogue, this new form takes place in a virtual forum that permits not only language but also visuals, many of which would be excluded from print publications. Second, the Internet is fostering a new form of collaboration in the creation of science and its visualization, one that allows amateurs to play a substantive role sometimes. Third, the Internet is fostering the creation and enabling the dissemination of visual representations more faithful to the Heideggerian vision of spatiotemporal magnitudes of motion, or, to translate Heidegger more literally, "space-time-motion magnitudes." Such visuals include 3-D images that rotate 360°, interactive images, and images that change in real or speeded-up time. A fourth byproduct of the expanded palette for visual representation online is a new form of virtual witnessing: for the first time, what readers experience can very nearly coincide with field and laboratory experience. Just as important as the new possibilities for scientific visualization on the Internet are innovative ways of presenting and arranging scientific content on the screen, with links to an extensive body of supplementary information.

Despite these considerable advantages, the present digital scientific article is still generally viewed conservatively as an electronically delivered stable network of words and pictures, data and concepts. The innovations we have surveyed are routinely seen as enhancements or deviations from the norm, not as major reconfigurations; indeed, many of the Internet visuals of the sort we have analyzed in this chapter, including the movies, are regularly labeled "supplementary." No scientific journal, no current means of scientific communication, realizes John Wilbanks's vision of the article as "inherently digital" (2006), not an object in real space, but a portal into the virtual space of the Internet. In this vision, the article accelerates scientific advance by providing access to semantic web-based networks and databases specific to fields or subfields. Employing such a system, scientists not only see a photograph, they also understand why it was taken, under what conditions, and with what equipment. Judgments can then be made concerning the quality of the image and its credibility. Such a system "would describe the actual contents of the image—what's in it, what it means, pointers to specific parts of the image, and so forth. This makes it possible to use the image scientifically, as well as to create links between images that share annotations and measurements and so forth" (Wilbanks, pers. comm., October 8, 2010; see also Neumann and Thomas 2002; Neumann, Miller, and Wilbanks 2004; Wilbanks 2006; Gudivada et al. 2008). What is true of photographs is true of all scientific images and graphics.

In our imagined cyberworld of the future, the complete shift to the inherently digital article will transform the experience of reading science. Readers will become nodes in a network of texts, pictures, persons, and Internet connections to immense databases; as a result of this alteration in role, they will have ready access to a vast warehouse of relevant information; at the same time, they will become aware of all possible collaborators, those working on the same or allied problems. This transformation in reading and work has another significant benefit. Because the number of viewers, citation counts, blog posts, trackbacks, and social bookmarking will be immediately accessible, all will have access to not only the content of articles but also the current community judgments of its value, a worth contingent also on new forms of peer review and new methods of self-regulation to certify credibility. In this new Internet world, a political value can also be realized for the first time: the democratization of knowledge. In more and more cases, credibility will be bestowed on the better argument, regardless of prestige and institutional power. Through the transformation of virtual witnessing, an empirical foundation for this better argument will also be in place. The German distinction, unavailable in English, is relevant to this new witnessing experience, the difference between *Erfahrung*, experience that makes its mark, and *Erlebnis*, experience we live through. By means of this transformation, scientist readers and authors can picture the same world, fully sharing the same epistemic perspective.

In this chapter, we have only scratched the surface of the Internet revolution, a turning point that has made it possible for verbal-visual interaction, so often at the center of science in the making, more and more to realize its creative and communicative potential, a horizon so presciently adumbrated in Heidegger's hermeneutics and his philosophy of science. The

Internet has also made it possible to renovate processes essential to scientific advance: collaboration, data collection and sharing, peer review, and post–peer review. In our next book, tentatively titled *The Future Is Already Here: The Internet Revolution in Science and Scholarship*, we will explore this combined impact at greater length and extend this exploration to the humanities, areas where this potential is beginning to be realized, opening up new areas of study, such as the history of musical performance, and new ways of studying established subjects, such as the history of the Civil War. As in the sciences, the Internet is also influencing scholarly collaboration, data collection and sharing, and peer review.

ACKNOWLEDGMENTS AND ENDNOTE

No man is an island, entire unto itself, and no printed or electronic book is either. For help in constructing this book, we wish to particularly thank librarians at the University of Minnesota Library, Lois Hendrickson and Elaine Challacombe, for assistance in tracking down books and reproducing some images. We would also thank four generous scientists, John Losos, Thomas Seeley, Matthew Meselson, Mario Capecchi for answering our queries; Christie Henry, editorial director at the University of Chicago Press, for her unwavering support over the years; the University of Chicago Press staff for handling the challenging job of producing a book with so many figures; and the University of Chicago Press readers for volunteering the time necessary to provide the constructive critical feedback all publishable scholarly books require. For inspiration, the second author (JEH) also thanks Marion Wood Covey, for whom "from sight to insight" is not just an occasional practice but a way of life. JEH also benefited immensely from many years of working with scientists in the creation and refinement of scientific images and tables of data as part of his job at Argonne National Laboratory. The work necessary to produce this book, however, was performed independently of Argonne and the US Department of Energy. Both authors would also like to thank Aydin Mohseni for his heroic efforts on behalf of our many images, and Anne Brataas for having introduced Aydin to us.

We are also greatly indebted to many authors and publishers for a rapid response to our request for permission to reproduce images in books, journal articles, PowerPoint presentations, and websites. Although we have contacted all copyright holders of images not in the public domain and have obtained licenses from those who wrote back asking for payment, we have not received answers from a few of them despite repeated inquiries. Nevertheless, we think that our employment of all of these images are instances of fair use in accordance to the "Code for Best Practices in Fair Use for Scholarly Research in Communication" created by the International Communication Association and endorsed by the National Communication Association of which the first author (AGG) is a member. The use of all of these images is transformative, each having been repurposed as an illustration in arguments we make for verbal-visual interaction in science, the thesis of this book. In no case have we used more material than was absolutely necessary for the arguments we make.

Finally, in several chapters, we reworked material that appeared previously in articles written by the first author (AGG): "The Verbal and the Visual in Science: A Heideggerian Perspective," *Science in Context* 19: 443–74 (2006); "Medical Tables, Graphics and Photographs: How They Work," *Journal of Technical Writing and Communication* 37, no. 4 (2007): 419–33; "The Brains in Brain: The Coevolution of Localization and Its Images," *Journal of the History of the Neurosciences* 17 (2008): 380–92; "Darwin's Diagram: Scientific Visions and Scientific Visuals," in *Ways of Seeing, Ways of Speaking: The Integration of Rhetoric and Vision in Constructing the Real*, ed. K. S. Fleckenstein, S. Hum, and L. T. Calendrillo (West Lafayette, IN: Parlor Press, 2007), 52–80. We thank the editors and reviewers of those publications for helping us sharpen our arguments.

REFERENCES

Ackerman, J. R., and R. W. Karrow, eds. 2007. *Maps: Finding Our Place in the World.* Chicago: University of Chicago Press.

Alley, M. A., and K. A. Neeley. 2005. Rethinking the design of presentation slides: A case for sentence headlines and visual evidence. *Technical Communication* 52:417–26.

Alley, M. A., M. Schreiber, K. Ramsdell, and J. Muffo. 2006. How the design of headlines in presentation slides affects audience retention. *Technical Communication* 53:225–34.

Aron, L., P. Klein, T.-T. Pham, E. R. Kramer, W. Wurst, and R. Klein. 2010. Pro-survival role for Parkinson's associated gene DJ-1 revealed in trophically impaired dopaminergic neurons. *PLoS Biology* 8 (4): e1000349.

Atkinson, Cliff. 2008. *Beyond Bullet Points: Using Microsoft Office PowerPoint 2007 to Create Presentations that Inform, Motivate, and Inspire.* Redmond, WA: Microsoft Press.

Bacon, Francis. (1627) 1915. *New Atlantis.* Edited by Alfred B. Gough. London: Oxford University Press.

Barrow, John D. 2008. *Cosmic Imagery: Key Images in the History of Science.* New York: W. W. Norton & Co.

Bartalena, L., C. Marcocci, F. Bogazzi, L. Manetti, M. L. Tanda, E. Dell'Unto, G. Bruno-Bossio, M. Nardi, M. P. Bartolomei, A. Lepri, G. Rossi, E. Martino, and A. Pinchera. 1998. Relation between therapy for hyperthyroidism and the course of Graves' ophthalmopathy. *New England Journal of Medicine* 338:73–78.

Bastide, Françoise. 1990. The iconography of scientific texts: Principles of analysis. In *Representation in Scientific Practice*, edited by M. Lynch and S. Woolgar, 187–229. Cambridge, MA: MIT Press.

Bateson, William. 1894. *Materials for the Study of Variation Treated with Especial Regard to Discontinuity in the Origin of Species.* London: Macmillan.

Bazerman, Charles. 1988. *Shaping Written Knowledge: The Genre and Activity of the Experimental Article in Science.* Madison: University of Wisconsin Press.

Berryman, Alan A. 1992. The origins and evolution of predator-prey theory. *Ecology* 73:1530–35.

Bertin, Jacques. 1981. *Graphics and Graphic Information Processing*. Translated by W. J. Berg and P. Scott. Berlin: Walter de Gruyter.

———. 1983. *Semiology of Graphics: Diagrams, Networks, Maps*. Translated by W. J. Berg. Madison: University of Wisconsin Press.

Bolton, J. S. 1903. The functions of the frontal lobes. *Brain* 26:215–41.

Boyer, Carl. 1987. *The Rainbow: From Myth to Mathematics*. Princeton, NJ: Princeton University Press.

Boyle, Robert. 1660. *New Experiments Physico-mechanicall, Touching the Spring of the Air, and Its Effects, Made, for the Most Part, in a New Pneumatic Engine*. Oxford: H. Hall.

Bradley, Jean-Claude. 2009a. Crowdsourcing solubility using open notebook science. http://www.scivee.tv/node/12302.

———. 2009b. Our attempt to reproduce an oxidation by Na. http://usefulchem.blogspot .com/2009/08/our-attempt-to-reproduce-oxidation-by.html.

Browne, Janet. 1985. Darwin and the expression of the emotions. In *The Darwinian Heritage*, edited by D. Kohn, 307–26. Princeton, NJ: Princeton University Press.

———. 1995. *Charles Darwin: Voyaging*. Princeton, NJ: Princeton University Press.

———. 2002. *Charles Darwin: The Power of Place*. New York: Alfred A. Knopf.

Brownstein, J. S., C. C. Freifeld, and L. C. Madoff. 2009. Digital disease detection— Harnessing the web for public health surveillance. *New England Journal of Medicine* 360:2153–57.

Brownstein, J. S., C. C. Freifeld, E. H. Chan, M. Keller, A. L. Sonricker, S. R. Mekaru, and D. L. Buckeridge. 2010. Information technology and global surveillance of cases of 2009 H1N1 Influenza. *New England Journal of Medicine* 362: 1731–35.

Burchfield, Joe D. 1990. *Lord Kelvin and the Age of the Earth*. Chicago: University of Chicago Press.

Capecchi, Mario R. 2007. Gene targeting into the 21st century. PowerPoint presentation for Nobel Lecture, December 7. http://nobelprize.org/nobel_prizes/medicine /laureates/2007/capecchi-lecture.html.

Carozzi, Albert V. 1965. Lavoisier's fundamental contribution to stratigraphy. *Ohio Journal of Science* 65:71–85.

Carpenter, P. A., and P. Shah. 1998. A model of the perceptual and conceptual processes in graph comprehension. *Journal of Experimental Psychology: Applied* 4:75–100.

Carr, Nicholas. 2010. *The Shallows: What the Internet Is Doing to Our Brains*. New York: W. W. Norton.

Carswell, C. B., J. R. Bates, N. R. Pregliasco, A. Lonon, and J. Urban. 1998. Finding graphs useful: Linking preference to performance for one cognitive tool. *Cognitive Technology* 3:4–18.

Cavina-Pratesi , C., M. A. Goodale, and J. C. Culham. 2007. FMRI reveals a dissociation between grasping and perceiving the size of real 3D objects. *PLoS ONE* 2 (5): e424.

Ceccarelli, Leah. 2001. *Shaping Science with Rhetoric: The Cases of Dobzhansky, Schrödinger, and Wilson*. Chicago: University of Chicago Press.

Chandrasekaran, B., and N. H. Narayanan. 1993. Perceptual representation and reasoning. In *Intelligent Systems: Concepts and Applications*, edited by L. S. Sterling, 67–76. New York: Plenum Press.

Clark, J. M., and A. Paivio. 1991. Dual coding theory and education. *Educational Psychology Review* 3:149–210.

Cleveland, W. S., and R. McGill. 1984. Graphical perception: Theory, experimentation, and application to the development of graphical methods. *Journal of the American Statistical Association* 79:531–54.

———. 1985. Graphical perception and graphical methods for analyzing scientific data. *Science* 229:828–33.

Cohen, C. 1998. Charles Lyell and the evidences of the antiquity of man. In *Lyell: The Past is Key to the Present*. Geological Society of London, Special Publications, edited by D. J. Bundell and A. C. Scott, vol. 143, 83–93.

Cohen, I. Bernard. 1985. *The Birth of the New Physics*. New York: W. W. Norton.

Cohen, M. R., and E. Nagel. 1934. *An Introduction to Logic and Scientific Method*. New York: Harcourt, Brace, and Company.

Collins, Harry. 2004. *Gravity's Shadow: The Search for Gravitational Waves*. Chicago: University of Chicago Press.

Collins, Randall. 1998. *The Sociology of Philosophies: A Global Theory of Intellectual Change*. Cambridge, MA: Harvard University Press.

Conard, N. J., M. Malina, and S. C. Münzel. 2009. New flutes document the earliest music tradition in southwestern Germany. *Nature* 460:737–40.

Cooper, S., F. Khatib, A. Treuille, J. Barbero, J. Lee, M. Beenen, A. Leaver-Fay, D. Baker, Z. Popović, and Foldit players. 2010. Predicting protein structures with a multiplayer online game. *Nature* 466:756–60.

Copernicus, Nicolas. (1543) 1978. *On the Revolutions*. Translated by Edward Rosen. London: Macmillan.

Coull, B. C. 1972. Species diversity and faunal affinities of Meiobenthic Copedoda in the deep sea. *Marine Biology* 14:48–51.

Cox, Allan. 1969. Geomagnetic reversals. *Science* 163:237–45.

———, ed. 1973. *Plate Tectonics and Geomagnetic Reversals*. San Francisco: W. H. Freeman.

Crick, Francis. 1988. *What Mad Pursuit: A Personal View of Scientific Discovery*. New York: Basic Books.

Crombie, Alistar C. 1959. *Medieval and Early Modern Science*. 2nd ed. Vol 2. New York: Doubleday.

Crosland, Maurice. 1978. *Historical Studies in the Language of Chemistry*. New York: Dover.

Dalton, John. 1808. *A New System of Chemical Philosophy*. Manchester: Printed by S. Russell for R. Bickerstaff.

Darwin, Charles. 1835. Darwin Correspondence Project Database, letter 275. http://www.darwinproject.ac.uk/entry-275/.

———. 1836. Geological notes made during a survey of the east and west coasts of S. America, in the years 1832, 1833, 1834 and 1835, with an account of a transverse section of the Cordilleras of the Andes between Valparaiso and Mendoza [Read by A. Sedgwick 18 November 1835]. *Proceedings of the Geological Society* 2:210–12.

———. 1837. On certain areas of elevation and subsidence in the Pacific and Indian

Oceans, as deduced from the study of coral formations [Read 31 May]. *Proceedings of the Geological Society of London* 2:552–54.

———. 1838. A paper, on the connexion of certain volcanic phaenomina, and on the formation of mountain-chains and volcanos, as effects of continental elevations [Read 7 March]. *Proceedings of the Geological Society of London* 2:654–60.

———. 1839. Observations on the parallel roads of Glen Roy, and of other parts of Lochaber in Scotland, with an attempt to prove that they are of marine origin [Read 7 February]. *Philosophical Transactions of the Royal Society* 129:39–81.

———. 1840. On the connexion of certain volcanic phenomena in South America; and on the formation of mountain chains and volcanos, as the effect of the same powers by which continents are elevated [Read 7 March 1838]. *Transactions of the Geological Society of London* 5 (2nd ser., part 3): 601–31.

———. 1842. *Structure and Distribution of Coral Reefs, Being the First Part of the Geology of the Voyage of the* Beagle *under the Command of Captain Fitzroy, R. N., during the Years 1832 to 1836.* London: Smith, Elder.

———. 1844. *Geological Observations on the Volcanic Islands Visited during the Voyage of* Beagle, *Together with Some Brief Notices on the Geology of Australia and the Cape of Good Hope, Being the Second Part of the Geology of the Voyage of the* Beagle, *under the Command of Capt. Fitzroy, R. N., during the Years 1832 to 1836.* London: Smith, Elder.

———. 1845. *Journal of Researches into the Natural History and Geology of the Countries Visited during the Voyage of HMS Beagle Round the World.* 2nd ed. London: John Murray.

———. 1846. *Geological Observations on South America, Being the Third Part of the Geology of the Voyage of the* Beagle, *under the Command of Capt. Fitzroy, R. N., during the years 1832 to 1836.* London: Smith, Elder.

———. (1859) 1964. *On the Origin of Species.* Cambridge, MA: Harvard University Press.

———. (1871) 1981. *The Descent of Man, and Selection in Relation to Sex.* 2 vols. Princeton, NJ: Princeton University Press.

———. (1872) 1965. *The Expression of the Emotions in Man and Animals.* Chicago: University of Chicago Press.

———. (1887) 1958. *The Autobiography of Charles Darwin, 1809–1882.* Edited by N. Barlow. London: Collins.

———. (1892) 1959. *The Life and Letters of Charles Darwin.* Edited by Francis Darwin. New York: Basic Books.

———. 1958. *The Autobiography of Charles Darwin and Selected Letters.* Edited by Francis Darwin. New York: Dover.

———. 1987a. *Charles Darwin's Notebooks.* Edited by P. H. Barrett, P. J. Gautrey, S. Herbert, D. Kohn, and S. Smith. Ithaca, NY: Cornell University Press.

———. 1987b. *The Correspondence, 1837–43.* Vol. 2. Edited by F. Burkhardt and S. Smith. Cambridge: Cambridge University Press.

Darwin, Francis. 1914. Francis Galton: 1822–1911. *Eugenics Review* 6:1–17.

Daston, L., and P. Galison. 2010. *Objectivity.* New York: Zone Books.

Dauenhauer, Bernard. 2005. Paul Ricoeur. In *Stanford Encyclopedia of Philosophy.* http://plato.stanford.edu/entries/ricoeur.

Dawkins, W. Boyd. 1874. *Cave Hunting*. London: Macmillan.

———. 1880. *Early Man in Britain and His Place in the Tertiary Period*. London: Macmillan.

Deledalle, Gérard. 2000. *Charles S. Peirce's Philosophy of Signs: Essays in Comparative Literature*. Bloomington: Indiana University Press.

D'Errico, F., C. Henshilwood, M. Vanhaeren, and K. van Niekerk. 2005. *Nassarius kraussianus* shell beads from Blombos Cave: Evidence for symbolic behaviour in the Middle Stone Age. *Journal of Human Evolution* 48:3–24.

Desnoyers, Luc. 2011. Toward a taxonomy of visuals in scientific communication. *Technical Communication* 58:119–34.

Dietz, R. S. 1961. Continent and ocean basin evolution by spreading of the sea floor. *Nature* 190:854–57.

Dijksterhuis, E. J. 1961. *The Mechanization of the World Picture*. Translated by C. Diksdhoorn. London: Oxford University Press.

Dobbs, David. 2005. Fact or phrenology? *Scientific American Mind* 16: 24–32.

Docherty, Paul. 2009. NaH as an oxidant—liveblogging! *Totally Synthetic*. http://totallysynthetic.com/blog/?p=1903.

Douglas, G. V., and A. V. Douglas. 1923. Note on the interpretation of the Wegener frequency curve. *Geological Magazine* 60:108–11.

Doumont, Jean-Luc. 2002. The three laws of professional communication. *IEEE Transactions on Professional Communication* 45:291–96.

Drake, Stillman. 1972. The uniform motion equivalent to a uniformly accelerated motion from rest. *Isis* 63:28–38.

Dubin, Mark W. 2002. *How the Brain Works*. Malden, MA: Blackwell Science.

Dumit, Joseph. 2004. *Picturing Personhood: Brain Scans and Biomedical Identity*. Princeton, NJ: Princeton University Press.

Durant, John D. 1985. The ascent of nature in Darwin's *Descent of Man*. In *The Darwinian Heritage*, edited by D. Kohn, 283–306. Princeton, NJ: Princeton University Press.

Dymock, S. 2007. Comprehension strategy instruction: Teaching narrative text structure awareness. *The Reading Teacher* 61:161–67.

Dziewonski, A. M., and J. H. Woodhouse. 1987. Global images of the earth's interior. *Science* 236:37–48.

Dzurisin, Daniel. 2006. The ongoing, mind-blowing eruption of Mount St. Helens. PowerPoint presentation, meeting of Geological Society of Oregon County, Portland State University, November 17. http://www.press.uchicago.edu/books/harmon/Harmon_PPT_example1_Dzurisin_Mt_St_Helens.ppt.

Einstein, Albert. 1959. Autobiographical notes. In *Albert Einstein: Philosopher-Scientist*, vol. 1, edited by Paul Arthur Schilp. New York: Harper and Row.

Eisenstein, Elizabeth. 1979. *The Printing Press as an Agent of Change*. Cambridge: Cambridge University Press.

Eisner, Will. 1994. *Comics and Sequential Art: Principles and Practice of the World's Most Popular Art Form*. Paramus, NJ: Poorhouse Press.

Evans, John. 1872. *Ancient Stone Implements, Weapons, and Ornaments of Great Britain*. London: Longmans, Green, Reader, and Dyer.

Fahnestock, Jeanne. 1999. *Rhetorical Figures in Science*. Oxford: Oxford University Press.

Faraday, Michael. 1821 (1952). On some new electro-magnetical motions and on the theory of magnetism. In *Great Books of the Western World*. Vol. 45, *Lavoisier, Fourier, Faraday*, edited by R. M. Hutchins, 795–809. Chicago: Encyclopaedia Britannica.

Farkas, David K. 2006. Toward a better understanding of PowerPoint deck design. *Information Design Journal and Document Design* 14:162–71.

Faye, Emmanuel. 2009. *Heidegger: The Introduction of Nazism into Philosophy in Light of the Unpublished Seminars of 1933–1935*. Translated by Michael B. Smith. New Haven, CT: Yale University Press.

Feynman, Richard. 1965. *The Character of Physical Law*. Cambridge, MA: MIT Press.

Fish, Menachem. 1985. Whewell's consilience of inductions: An evaluation. *Philosophy of Science* 52:239–55.

Fisher, R. A., and E. B. Ford. 1947. The spread of a gene in natural conditions in a colony of the moth, *Panaxia dominula*, L. *Heredity* 1:143–74.

Fleck, Ludwik. 1979. *Genesis and Development of a Scientific Fact*. Translated by T. J. Trenn and R. K. Merton. Chicago: University of Chicago Press.

Fleischacker, Samuel. 2008. *Heidegger's Jewish Followers: Essays on Hannah Arendt, Leo Strauss, Hans Jonas, and Emmanuel Levinas*. Pittsburgh: Duquesne University Press.

Fogelin, Lars. 2007. Inference to the best explanation: A common and effective form of archaeological explanation. *American Antiquity* 72:603–25.

Forster, Malcolm R. 1988. The confirmation of common component causes. *PSA: Proceedings of the Biennial Meeting of the Philosophy of Science Association* 1:3–9.

Frankel, Felice. 2002. *Envisioning Science: The Design and Craft of the Scientific Image*. Cambridge, MA: MIT Press.

Frankel, Henry. 1976. Alfred Wegener and the specialists. *Centaurus* 20:305–24.

———. 1987. The continental drift debate. In *Scientific Controversies: Case Studies in the Resolution and Closure of Disputes in Science and Technology*, edited by H. T. Engelhardt Jr. and A. L. Caplan, 203–48. Cambridge: Cambridge University Press.

Freedberg, David. 2002. *The Eye of the Lynx: Galileo, His Friends, and the Beginnings of Modern Natural History*. Chicago: University of Chicago Press.

Friedland, G. W., and M. Friedman. 1998. *William Roentgen and the X-Ray Beam: Medicine's Ten Greatest Discoveries*. New Haven, CT: Yale University Press.

Friedman, Michael. 2000. *A Parting of the Ways: Carnap, Cassirer, and Heidegger*. Chicago: Open Court.

Friel, S. N., F. R. Curcio, and G. W. Bright. 2001. Making sense of graphs: Critical factors influencing comprehension and instructional implications. *Journal for Research in Mathematics Education* 32 (2):124–58.

Friendly, M., and D. J. Denis. 2010. Milestones in the history of thematic cartography, statistical graphics, and data visualization. http://datavis.ca/milestones.

Galilei, Galileo. (1610) 1880. *The Sidereal Messenger*. Translated by Edward Stafford Carlos. London: Rivington.

———. (1632) 1967. *Dialogue Concerning the Two Chief World Systems*. Translated by Stillman Drake. Berkeley: University of California Press.

———. (1638) 1974. *Two New Sciences*. Translated by Stillman Drake. Madison: University of Wisconsin Press.

Galison, Peter. 1997. *Image and Logic: A Material Culture of Microphysics.* Chicago: University of Chicago Press.

———. 1998. Judgment against objectivity. In *Picturing Science, Producing Art*, edited by C. A. Jones and P. Galison, 327–59. New York: Routledge.

———. 1999. Objectivity is romantic. Occasional paper 47, American Council of Learned Societies.

———. 2008. Ten problems in history and philosophy of science. *Isis* 99:111–24.

Galpin, A. J., and G. Underwood. 2005. Eye movements during search and detection in comparative visual search. *Attention, Perception & Psychophysics* 67:1313–31.

Garner, J. K., M. Alley, A. F. Gaudelli, and S. E. Zappe. 2009. Common use of PowerPoint versus assertion-evidence slide structure: A cognitive psychology perspective. *Technical Communication* 56:331–45.

Geikie, James. 1881. *Prehistoric Europe: A Geological Sketch.* London: Edward Stanford.

———. 1894. *The Great Ice Age and Its Relation to the Antiquity of Man.* 3rd ed. London: Edward Stanford.

Geoffroy, Étienne-François. 1718. Table des différents rapports observés en chemie entre différents substances. *Mémoires de l'Académie Royale des Sciences*, 202–12.

Giere, Ronald N. 1996. Visual models and scientific judgment. In *Picturing Knowledge: Historical and Philosophical Problems Concerning the Use of Art in Science*, edited by Brian S. Baigre. Toronto: University of Toronto Press.

Gladwell, Malcolm. 2010. The treatment. *New Yorker*, May 17, 69.

Glazebrook, Trish. 2000. *Heidegger's Philosophy of Science.* New York: Fordham University Press.

Goodman, Nelson. 1972. *Problems and Prospects.* Indianapolis: The Bobbs-Merrill Company.

Goodsell, D. S. 2005. Visual methods from atoms to cells. *Structure* 13:347–54.

Gould, Stephen Jay. 1998. Capturing the center: Antoine-Laurent Lavoisier's scientific contributions. *Natural History* 107 (December): 14–25.

Grayson, D. K. 1983. *The Establishment of Human Antiquity.* New York: Academic Press.

Griesemer, J. R., and W. C. Wimsatt. 1989. Picturing Weismannism: A case study of conceptual evolution. In *What the Philosophy of Biology Is*, edited by Michael Ruse, 75–138. Dordrecht: Kluwer Academic Publishers.

Gross, Alan G. 1998. Do disputes over priority tell us anything about science? *Science in Context* 11:161–79.

———. 2006. *Starring the Text: The Place of Rhetoric in Science Studies.* Carbondale: Southern Illinois University Press.

Gross, A. G., J. E. Harmon, and M. Reidy. 2002. *Communicating Science: The Scientific Article from the Seventeenth Century to the Present.* Oxford: Oxford University Press. Reprinted by Parlor Press, 2009.

Gudivada, R. C., X. A. Qu, J. Chen, A. G. Jegga, E. K. Neumann, and B. J. Aronow. 2008. Identifying disease-causal genes using semantic web-based representation of integrated genomic and phenomic knowledge. *Journal of Biomedical Informatics* 41:717–29.

Habermas, Jürgen. 1987. *The Theory of Communicative Action.* Vol. 2. Translated by T. McCarthy. Boston: Beacon Press.

Habermas, Tilmann. 2007. How to tell a life: The development of the cultural concept of biography. *Journal of Cognition and Development* 8:1–31.

Hammer, N. J., H.-Th. Janka, and E. Müller. 2010. Three-dimensional simulations of mixing instabilities in supernova explosions. *Astrophysical Journal* 714:1371–85.

Hand, Eric. 2010. People power: Networks of human minds are taking citizen science to a new level. *Nature* 466:685–87.

Hankins, Thomas L. 1999. Blood, dirt, and nomograms: A particular history of graphs. *Isis* 90:50–80.

Hansen, D. V., J. H. Lui, P. R. L. Parker, and A. R. Kriegstein. 2010. Neurogenic radial glia in the outer subventricular zone of human neocortex. *Nature* 464:554–61.

Hanson, Norwood R. 1958. *Patterns of Discovery: An Inquiry into the Conceptual Foundations of Science.* Cambridge: Cambridge University Press.

Hariman, R., and J. L. Lucaites. 2002. Performing civic identity: The iconic photograph of the flag raising on Iwo Jima. *Quarterly Journal of Speech* 88:363–92.

Harnad, Stevan. 1991. Post-Gutenberg galaxy: The fourth revolution in the means of production of knowledge. *Public-Access Computer Systems Review* 2:39–53.

Hayes, Brian. 1998. The invention of the genetic code. *American Scientist* 86:8–14.

Hayles, N. Katherine. 2008. *Electronic Literature: New Horizons for the Literary.* Notre Dame, IN: University of Notre Dame Press.

Heelan, Patrick A. 1983. *Space-Perception and the Philosophy of Science.* Berkeley: University of California Press.

Heidegger, Martin. (1927–28) 1977. *Phenomenological Interpretation of Kant's "Critique of Pure Reason."* Translated by P. Emad and K. Maly. Bloomington: Indiana University Press.

———. (1935–36) 1967. *What Is a Thing?* translated by W. B. Barton, Jr., and V. Deutsch. Chicago: Henry Regnery Company.

———. (1938) 1977. The [era] of the world picture. In *The Question Concerning Technology and Other Essays,* translated by W. Lovett, 115–54. New York: Harper and Row.

———. 1950. Die Zeit des Weltbildes. In *Holzwege,* 69–104. Frankfurt am Main: Vittorio Klostermann.

———. (1953a) 1977. The question concerning technology. In *The Question Concerning Technology and Other Essays,* translated by W. Lovett, 3–35. New York: Harper and Row.

———. (1953b) 1977. Science and reflection. In *The Question Concerning Technology and Other Essays,* translated by W. Lovett, 155–82. New York: Harper and Row.

———. 1954a. Wissenschaft und Besinnung. In *Vorträger und Aufsätze,* 45–70. Pfullingen: Gunther Ness.

———. 1954b. Die Frage Nach Der Technik. In *Vorträger und Aufsätze,* 13–44. Pfullingen: Gunther Ness.

———. 1962a. *Being and Time.* Translated by J. Macquarrie & E. Robinson. New York: Harper and Row.

———. 1962b. *Die Frage nach dem Ding.* Tübingen: Max Neimeyer.

———. 1972. *Sein und Zeit,* 12th edn. Tübingen: Max Niemeyer Verlag.

Hempel, Carl G. 1965. *Aspects of Scientific Explanation and Other Essays in the Philosophy of Science.* New York: The Free Press.

Henshilwood, C., F. d'Errico, M. Vanhaeren, K. van Niekerk, and Z. Jacobs. 2004. Middle Stone Age shell beads from South Africa. *Science* 304:404.

Higham, T., L. Basell, R. Jacobic, R. Wooda, C. B. Ramseya, and N. J. Conard. 2012. Testing models for the beginnings of the Aurignacian and the advent of figurative art and music: The radiocarbon chronology of Geißenklösterle. *Journal of Human Evolution* 62(6): 664–76.

His, Wilhelm. 1880. *Anatomie menschlicher Embryonen. I. Embryonen des ersten Monats.* Leipzig: F. C. W. Vogel.

Hobson, J. A., and R. W. McCarley. 1977. The brain as a dream state generator: An activation-synthesis hypothesis of the dream process. *American Journal of Psychiatry* 134:1335–48.

Hochberg, Julian. 1998. Gestalt theory and its legacy: Organization in eye and brain, in attention and mental representation. In *Perception and Cognition at Century's End,* edited by J. Hochberg, 253–306. San Diego: Academy Press.

Hoffmann, Roald. 1981. Theoretical organometallic chemistry. *Science* 211:995–1002.

———. 1991. Art in Science? *Q (A Journal of Art),* May, 62–65.

Holden, Constance. 2004. Oldest beads suggest early symbolic behavior. *Science* 304:369.

Holmes, Frederic Lawrence. 2001. *Meselson, Stahl, and the Replication of DNA: A History of "The Most Beautiful Experiment in Biology."* New Haven, CT: Yale University Press.

Holy, T. E., and Z. Guo. 2005. Ultrasonic songs of male mice. *PLoS Biology* 3:2177–86.

Hubble, Edwin. 1929. A relation between distance and radial velocity among extragalactic nebulae. *Proceedings of the National Academy of Sciences* 15:168–73.

Hull, David. 1973. *Darwin and His Critics: The Reception of Darwin's Theory of Evolution by the Scientific Community.* Chicago: University of Chicago Press.

Hume, David. (1748) 1955. *An Inquiry Concerning Human Understanding.* Indianapolis: Bobbs-Merrill.

Humphrey, M. E., and O. L. Zangwill. 1952. Effects of a right-sided occipito-parietal brain injury on a left-handed man. *Brain* 75:312–24.

Ihde, Don. 1991. *Instrumental Realism: The Interface between Philosophy of Science and Philosophy of Technology.* Bloomington: Indiana University Press.

———. 1998. *Expanding Hermeneutics: Visualism in Science.* Evanston: Northwestern University Press.

Isacks, B., J. Oliver, and L. R. Sykes. 1968. Seismology and the new global tectonics. *Journal of Geophysical Research* 73:5855–99.

Iverson, Richard M. 2006. A dynamic model of seismogenic volcanic extrusion, Mount St. Helens, 2004–2005. PowerPoint presentation at fall meeting, American Geophysical Society, December 5–9, http://www.press.uchicago.edu/books/harmon/Harmon_PPT_example3_Iverson_Mt_St_Helens.ppt.

Iverson, R. M., D. Dzurisin, C. A. Gardner, T. M. Gerlach, R. G. LaHusen, M. Lisowski, J. J. Major, S. D. Malone, J. A. Messerich, S. C. Moran, J. S. Pallister, A. I. Qamar, S. P. Schilling, and J. W. Vallance. 2006. Dynamics of seismogenic volcanic extrusion at Mount St. Helens in 2004–05. *Nature* 444:439–43.

Ivins, William M., Jr. 1969. *Prints and Visual Communication.* Cambridge, MA: MIT Press.

James, T. W., J. Culham, C. K. Humphrey, A. D. Milner, and M. A. Goodale. 2003. Ventral occipital lesions impair object recognition but not object-directed grasping. *Brain* 126:2463–75.

Janis, I. L., and L. Mann. 1977. *Decision Making: A Psychological Analysis of Conflict, Choice, and Commitment.* New York: Free Press.

Jones, C. A., and P. Galison, eds. 1998. *Picturing Science, Producing Art.* New York: Routledge.

Kaiser, David. 2005. *Drawing Theories Apart: The Dispersion of Feynman Diagrams in Postwar Physics.* Chicago: University of Chicago Press.

Kant, Immanuel. (1787) 1965. *Critique of Pure Reason.* Translated by N. K. Smith. New York: St. Martin's Press.

Karlberg, T., S. van den Berg, M. Hammarström , J. Sagemark, I. Johansson I, L. Holmberg-Schiavone, and H. Schüler. 2010. Crystal structure of the ATPase domain of the human AAA+ protein paraplegin/SPG7. *PLoS ONE* 4 (10): e6975.

Karrow, Robert W. 2007. Introduction. In *Maps: Finding Our Place in the World,* edited by J. R. Ackerman and R. W. Karrow, 1–17. Chicago: University of Chicago Press.

Keller, Corey, ed. 2008. *Brought to Light: Photography and the Invisible, 1840–1900.* San Francisco Museum of Modern Art.

Kemp, Martin. 2000. *Visualizations: The Nature Book of Art and Science.* Berkeley: University of California Press.

Kirby, K. N., and S. M. Kosslyn. 1992. Thinking visually. In *Understanding Vision: An Interdisciplinary Perspective,* edited by G. W. Humphreys, 71–86. Oxford: Blackwell.

Knorr-Cetina, Karen D. 1981. *The Manufacture of Knowledge: An Essay on the Constructivist and Contextual Nature of Science.* Oxford: Pergamon Press.

———. 1999. *Epistemic Cultures: How the Sciences Make Knowledge.* Cambridge, MA: Harvard University Press.

Køber, L., P. E. B. Thomsen, M. Møller, C. Torp-Pedersen, J. Carlsen, E. Sandøe, K. Egstrup, E. Agner, J. Videbæk, and B. Marchant. 2000. Effect of dofetilide in patients with recent myocardial infarction and left-ventricular dysfunction: A randomized trial. *Lancet* 356:2052–58.

Köhler, Wolfgang. 1947. *Gestalt Psychology: An Introduction to New Concepts in Modern Psychology.* New York: New American Library.

Kosslyn, Stephen M. 1989. Understanding charts and graphs. *Applied Cognitive Psychology* 3:185–225.

———. 2002. Einstein's mental images: The role of visual, spatial, and motoric representations. In *The Languages of the Brain,* edited by A. M. Galaburda, S. M. Kosslyn, and Y. Christen, 271–87. Cambridge, MA: Harvard University Press.

———. 2006. *Graph Design for the Eye and Mind.* Oxford: Oxford University Press.

———. 2007. *Clear and to the Point: Eight Psychological Principles for Compelling PowerPoint Presentations.* New York: Oxford University Press.

Kress, G., and T. van Leeuwen. 1996. *Reading Images: The Grammar of Visual Design.* London: Routledge.

———. 2001. *Multimodal Discourse: The Modes and Media of Contemporary Communication*. London: Arnold.

Krishnan, J. A., and R. A. Mularski. 2010. Acting on comparative effectiveness research in COPD. *Journal of the American Medical Association* 303:2409–10.

Kruggel, F., J. Turner, L. Tugan Muftuler, and The Alzheimer's Disease Neuroimaging Initiative. 2010. The impact of scanner hardware and imaging protocol on image quality and compartment volume precision in the ADNI cohort. *Neuroimage* 49:2123–33.

Langmuir, Irving. 1989. Pathological science. *Physics Today* 42:36–48.

Larkin, J. H., and H. A. Simon. 1987. Why a diagram is (sometimes) worth ten thousand words. *Cognitive Science* 11:65–99.

Latour, Bruno. 1986. Visualization and cognition: Thinking with eyes and hands. *Knowledge and Society* 6:1–40.

———. 1987. *Science in Action: How to Follow Scientists and Engineers through Society*. Cambridge, MA: Harvard University Press.

———. 1990. Drawing things together. In *Representation in Scientific Practice*, edited by M. Lynch and S. Woolgar, 19–68. Cambridge, MA: MIT Press.

Latour, B., and S. Woolgar. 1979. *Laboratory Life: The Construction of Scientific Facts*. Princeton, NJ: Princeton University Press.

Laudan, Larry. 1971. William Whewell on the consilience of inductions. *Monist* 55:368–91.

Laudan, Rachel. 1980. The method of multiple working hypotheses and the development of plate tectonic theory. In *Scientific Discovery: Case Studies*, edited by Thomas Nickles, 331–43. Dordrecht: D. Reidel.

Lavoisier, Antoine. 1789a. Observations générales sur les couches modernes horizontales qui ont été déposées par la mer et sur les consequences qu'on peut tirer de leurs dispositions relativement à l'ancienneté du globe terrestre. *Mémoires de l'Academié Royale des Sciences*, 186–204.

———. 1789b. *Traité élémentaire de chimie, présenté dans un ordre nouveau, et d'après les découvertes modernes*. Paris: Cuchet.

———. 1892. *Oeuvres de Lavoisier*, vol. 5. Paris: Imprimerie Nationale.

LeGrand, Homer E. 1988. *Drifting Continents and Shifting Theories*. Cambridge: Cambridge University Press.

Levmore, S., and M. C. Nussbaum. 2011. *Offensive Speech: Speech, Privacy, and Reputation*. Cambridge, MA: Harvard University Press.

Lewis, Cherry. 2000. *The Dating Game: One Man's Search for the Age of the Earth*. Cambridge: Cambridge University Press.

Lindberg, David C. 1978. The science of optics. In *Science in the Middle Ages*, edited by David C. Lindberg, 338–68. Chicago: University of Chicago Press.

Lindenauer, P. K., P. S. Pekow, M. C. Lahti, Y. Lee, E. M. Benjamin, and M. B. Rothberg. 2010. Association of corticosteroid dose and route of administration with risk of treatment failure in acute exacerbation of chronic obstructive pulmonary disease [COPD]. *Journal of the American Medical Association* 303:2359–67.

Linnaeus, Carl. (1735) 1964. *Systema Naturae*. Translated by M. S. J. Engel-Ledeboer and H. Engel, 19. Nieuwkoop: De Graaf.

Lipton, Peter. 2004. *Inference to the Best Explanation*. 2nd ed. London: Routledge.

Liu, C., Z. Dutton, C. H. Behroozi, and L. V. Hau. 2001. Observation of coherent optical information storage in an atomic medium using halted light pulses. *Nature* 409: 490–93.

Losos, Jonathan B. 2009. Integrating ecological and evolutionary studies of biological diversity: Patterns and processes of adaptive radiation in Caribbean lizards. PowerPoint presentation.

———. 2010. Adaptive radiation, ecological opportunity, and evolutionary determinism. *American Naturalist* 175:623–39.

Losos, J. B., and R. E. Ricklefs. 2009. Adaptation and diversification on islands. *Nature* 457:830–36.

Losos, J. B., T. W. Schoener, and D. A. Spiller. 2004. Predator-induced behavior shifts and natural selection in field-experiment lizard populations. *Nature* 432:505–8.

Lubbock, John. 1875. *Pre-historic Times, as Illustrated by Ancient Remains, and the Manners and Customs of Modern Savages.* 2nd ed. New York: D. Appleton.

Lyell, Charles. (1832) 1997. *Principles of Geology.* London: Penguin Classics.

———. 1863. *The Geological Evidences of the Antiquity of Man with Remarks on Theories of the Origin of Species by Variation.* London: John Murray.

Lynch, Michael. 1985. *Art and Artifact in Laboratory Science: A Study of Shop Work and Shop Talk in a Research Laboratory.* Boston: Routledge and Kegan Paul.

———. 1990. The externalized retina: Selection and mathematization in the visual documentation of objects in the life sciences. In *Representation in Scientific Practice,* edited by M. Lynch and S. Woolgar, 153–86. Cambridge, MA: MIT Press.

Lynch, M., and S. Woolgar, eds. 1990a. *Representation in Scientific Practice.* Cambridge, MA: MIT Press.

———. 1990b. Introduction: Sociological orientations to representational practice in science. In *Representation in Scientific Practice,* edited by M. Lynch and S. Woolgar. Cambridge, MA: MIT Press.

MacEachren, Alan M. 2004 *How Maps Work: Representation, Visualization, and Design.* New York: Guilford Press.

Mackiewicz, Jo. 2005. Review of *Use and Effect of Declarative Information in User Instructions,* by J. Karreman. *IEEE Transactions of Professional Communication* 48:105–6.

Magnan, Valentin. 1878–79. General paralysis and cerebral tumour, with atrophy of the ascending parietal convolution of the left hemisphere—no paralysis on the right side—convulsions on the left. *Brain* 1:561–65.

Malinas, Gary. 2003. Truth, negation, and entailment in pictures. In *DMS: Proceedings of the Ninth International Conference on Distributed Multimedia Systems, Knowledge Systems Institute.* Section: *Proceedings of the 2003 Conference on Visual Languages and Computing,* edited by Alfonso Cardenas and Piero Mussio, 257–62. Miami: Florida International University.

Marcuvitz, Nathan. 1951. *Waveguide Handbook.* New York: McGraw-Hill.

Markel, Mike. 2009. Exploiting verbal-visual synergy in presentation slides. *Technical Communication* 56:122–31.

Mendeleev, Dimitri. 1869. On the relationship of the properties of the elements to their atomic weights. *Zhurnal Russkoe Fiziko-Khimicheskoe Obshchestvo* 1:60–77; abstracted in *Zeitschrift für Chemie* 12:405–6.

Mercier, H., and D. Sperber. 2011. Why do humans reason? Arguments for an argumentative theory. *Behavioral and Brain Sciences* 34:57–111.

Merton, Robert K. 1973. *The sociology of science: Theoretical and empirical investigations.* Chicago: University of Chicago Press.

Meselson, M., and F. W. Stahl. 1958. The replication of DNA in *Escherichia coli. Proceedings of the National Academy of Sciences,* 44:671–82.

Miller, Arthur I. 1984. *Imagery in Scientific Thought: Creating Twentieth-Century Physics.* Boston: Birkhäuser.

Milner, A. D., and M. A. Goodale. 2006. *The Visual Brain in Action.* 2nd ed. Oxford: Oxford University Press.

Montgomery, Scott L. 1996. *The Scientific Voice.* New York: Guilford Press.

Moss, Jean Dietz. 1993. *Novelties in the Heavens: Rhetoric and Science in the Copernican Controversy.* Chicago: University of Chicago Press.

Myers, Greg. 1990. *Writing Biology: Texts in the Social Construction of Scientific Knowledge.* Madison: University of Wisconsin Press.

Neeley, K. A., M. Alley, C. G. Nicometo, and L. C. Spajek. 2009. Challenging the common practice of PowerPoint at an institution: Lessons from instructors. *Technical Communication* 56:346–60.

Nehamas, A., and P. Woodruff. 1995. Introduction. In *Phaedrus by Plato.* Indianapolis: Hackett Publishing.

Neumann, E. K., E. Miller, and J. Wilbanks. 2004. What the semantic web could do for the life sciences. *Drug Development Today* 2:228–36.

Neumann, E., and J. Thomas. 2002. Knowledge assembly for the life sciences. *Drug Development Today* 7: S160–62.

Nickles, Thomas. 1980. Scientific discovery and the future of philosophy of science. In *Scientific Discovery, Logic, and Rationality,* edited by Thomas Nickles, 1–59. Dordrecht: D. Reidel.

Nielsen, Michael. 2011. *Reinventing Discovery: The New Era of Networked Science.* Princeton, NJ: Princeton University Press.

North, Alison J. 2006. Seeing is believing? A beginner's guide to practical pitfalls in image acquisition. *Journal of Cell Biology* 172 (1): 9–18.

Oestermeier, U., and F. W. Hesse. 2000. Verbal and visual causal arguments. *Cognition* 75:65–104.

Ojemann, G. A., J. G. Ojemann, E. Lettich, and M. Berger. 1989. Cortical language localization in the left, dominant hemisphere. *Journal of Neurosurgery* 71:316–26.

Ongerboer de Visser, B. W., and H. G. J. M. Kuypers. 1978. Late blink reflex changes in lateral medullary lesions: An electrophysiological and neuro-anatomical study of Wallenberg's syndrome. *Brain* 101:285–94.

Opdyke, N. D., B. Glass, J. D. Hays, and J. Foster. 1966. Paleomagnetic study of Antarctic deep-sea cores. *Science* 154:349–57.

Oreskes, Naomi. 1999. *The Rejection of Continental Drift: Theory and Method in American Earth Science.* New York: Oxford University Press.

Owen, John Stewart Mackenzie. 2005. *The Scientific Article in the Age of Digitization.* Amsterdam: Springer.

Paivio, Allan. 2007. *Mind and Its Evolution: A Dual Coding Theoretical Approach.* Mahwah, NJ: Lawrence Erlbaum.

Panese, Francesco. 2006. The accursed part of scientific iconography. In *Visual Cultures of Science: Rethinking Representational Practices in Knowledge Building and Science Communication,* edited by L. Pauwels, 63–89. Hanover, NH: Dartmouth College Press/University Press of New England.

Pauwels, Luc, ed. 2006. *Visual Cultures of Science: Rethinking Representational Practices in Knowledge Building and Science Communication.* Hanover, NH: Dartmouth College Press/University Press of New England.

Peirce, Charles Sanders. 1955. *Philosophical Writings of Peirce.* Edited by Justus Buchler. New York: Dover.

Pera, Marcello. 1994. *The Discourses of Science.* Translated by Clarissa Botford. Chicago: University of Chicago Press. Revision of *Scienza e retorica,* 1991.

Perini, Laura. 2005a. The truth in pictures. *Philosophy of Science* 72:262–85.

———. 2005b. Visual representations and confirmation. *Philosophy of Science* 72: 913–26.

Perrault, Claude. 1669. Description anatomique d'un cameleon, d'un castor, d'un dromadaire, d'un ours, et d'une gazelle. Paris: Frederic Leonard.

Perrin, Jean. 1909. Mouvement brownien et réalité moléculaire. *Annales de chimie et de physique* 18:5–114. Translation by Frederick Soddy, *Brownian Movement and Molecular Reality,* London: Taylor and Francis, 1910.

Phillips, J. D. 1974. The broad—and broadening—Atlantic. *Oceanus* 17:20–27.

Pinker, Steven. 1983. *Pattern Perception and the Comprehension of Graphs.* Eric Document Service no. ED 1.310/2.237339.

———. 1990. A theory of graph comprehension. In *Artificial Intelligence and the Future of Testing,* edited by R. Freedle, 73–126. Hillsdale, NJ: Lawrence Erlbaum.

Pitman, W. C., and J. R. Heirtzler. 1966. Magnetic anomalies over the Pacific-Antarctic ridge. *Science* 154:1164–71.

Plato. (360 BCE) 1956. *Phaedrus.* Translated by W. C. Helmbold and W. G. Rabinowitz. Indianapolis: Library of Liberal Arts/Bobbs.

———. (360 BCE) 1928. *The Seventh Letter.* Translated by J. Harward. http://classics.mit .edu/Plato/seventh_letter.html.

Price, J. R., J. E. Roberts, and S. C. Jackson. 2006. Structural development of the fictional narratives of African American preschoolers. *Language, Speech, and Hearing Services in Schools* 37:178–90.

Quirk, R., S. Greenbaum, G. Leech, and J. Svartvik. 1972. *A Grammar of Contemporary English.* London: Longman.

Rayner, Keith. 2009. Eye movements in reading: Models and data. *Journal of Eye Movement Research* 2 (5): 1–10.

Reichenbach, Hans. 1938. *Experience and Prediction: An Analysis of the Foundation and the Structure of Knowledge.* Chicago: University of Chicago Press.

Reiser, Stanley J., 1991. The clinical record in medicine. Part 1: Learning from cases. *Annals of Internal Medicine* 114: 902–7.

Reitsma, Ella. 2008. *Maria Sibylla Merian & Daughters: Women of Art and Science.* Los Angeles: J. Paul Getty Museum.

Roberts, Lissa. 1991. Setting the table: The disciplinary development of eighteenth-century chemistry read through the changing structure of its tables. In *The Literary Structure of Scientific Argument: Historical Studies,* edited by Peter Dear, 99–132. Philadelphia: University of Pennsylvania Press.

Rocke, Alan J. 2005. In search of El Dorado: John Dalton and the origins of the atomic theory. *Social Research* 72:125–58.

———. 2010. *Image and Reality: Kekulé, Kopp, and the Scientific Imagination.* Chicago: University of Chicago Press.

Rojcewicz, Richard. 2006. *The Gods and Technology: A Reading of Heidegger.* Albany: SUNY Press.

Röntgen, Wilhelm Konrad. 1895. Ueber eine neue Art von Strahlen. *Sitzungsberichte der Physikalische-medicinischen Gesellschaft zu Würzberg,* 137–47.

———. 1896. On a new kind of rays. Translated by Arthur Stanton. *Nature* 53:274–76.

Rorty, Amélie Oksenberg, ed. 1980. *Explaining Emotions.* Berkeley: University of California Press.

Rosch, Eleanor. 1973. Natural categories. *Cognitive Psychology* 4:328–50.

———. 1975. Cognitive reference points. *Cognitive Psychology* 7:532–47.

———. 1999. Reclaiming concepts. *Journal of Consciousness Studies* 6:61–77.

Rosch, E., and C. B. Mervis. 1975. Family resemblance: Studies in the internal structure of categories. *Cognitive Psychology* 7:573–605.

Rosch, E., C. B. Mervis, W. D. Gray, D. M. Johnson, and P. Boyes-Braem. 1976. Basic objects in natural categories. *Cognitive Psychology* 8:382–439.

Rosen, Brian. 1982. Darwin, coral reefs, and global energy. *BioScience* 32:519–25.

Rossner, M. and K. M. Yamada. 2004. What's in a picture? The temptation of image manipulation. *Journal of Cell Biology* 166:11–15.

Rudwick, Martin J. S. 1976. The emergence of a visual language for geological science, 1760–1840. *History of Science* 16:149–95.

———. 1985. *The Great Devonian Controversy: The Shaping of Scientific Knowledge among Gentlemanly Specialists.* Chicago: University of Chicago Press.

———. 2005. *Bursting the Limits of Time: The Reconstruction of Geohistory in the Age of Revolution.* Chicago: University of Chicago Press.

Ryle, Gilbert. (1949) 2000. *The Concept of Mind.* Chicago: University of Chicago Press.

Saussure, Ferdinand de. (1916) 1986. *Course in General Linguistics.* Edited by C. Bally, A. Sechehasye, and A. Riedlinger. Translated by R. Harris. La Salle, IL: Open Court.

Schickore, J., and F. Steinle, eds. 2006. *Revisiting Discovery and Justification: Historical and Philosophical Perspectives on the Context Distinction.* Dordrecht: Springer.

Schmidt, B. P., N. B. Suntzeff, M. M. Phillips, R. A. Schommer, A. Clocchiatti, R. P. Kirshner, P. Garnavich, P. Challis, B. Leibundgut, J. Spyromilio, A. G. Riess, A. V. Filippenko, M. Hamuy, R. C. Smith, C. Hogan, C. Stubbs, A. Diercks, D. Reiss, R. Gilliland, J. Tonry, J. Maza, A. Dressler, J. Walsh, and R. Ciardullo. 1998. The high-Z supernova search: Measuring cosmic deceleration and global curvature of the universe using Type Ia supernovae. *Astrophysical Journal* 507:46–63.

Schoener, T. W., D. A. Spiller, and J. Losos. 2004. Variable ecological effects of hurricanes: The importance of seasonal timing for survival of lizards on Bahamian islands. *Proceedings of National Academy of Sciences* 101:177–81.

Schoonover, Carl. 2010. *Portraits of the Mind: Visualizing the Brain from Antiquity to the 21st Century*. New York: Abrams.

Schwarzbach, Martin. 1986. *Alfred Wegener: The Father of Continental Drift*. Translated by Carla Love. Madison: Science Tech Inc.

Scrope, George. 1827. *Memoir on the Geology of Central France*. London: Longman, Rees, Orme, Brown, and Green.

Sebeok, Thomas A. 1976. Iconicity. *MLN* 9:1427–56.

Secord, James A. 2000. *Victorian Sensations: The Extraordinary Publication, Reception, and Secret Authorship of "Vestiges of the Natural History of Creation."* Berkeley: University of California Press.

Seeley, Thomas D. 2005. House hunting by honey bees: A study in group decision-making. PowerPoint presentation at Cornell University, Department of Entomology, April 1. http://www.press.uchicago.edu/books/harmon/Harmon_PPT_example2 _House_Hunting.ppt.

Seeley, T. D., P. K. Visscher, and K. M. Passino. 2006. Group decision making in honey bee swarms. *American Scientist* 94:220–29.

Shaikh, N., A. Hoberman, P. H. Kaleida, D. L. Ploof, and J. L. Paradise. 2010. Diagnosing otitis media: Otoscopy and cerumen removal. *New England Journal of Medicine* 362: e62.

Shapin, S., and S. Schaffer. 1985. *Leviathan and the Air-Pump: Hobbes, Boyle, and the Experimental Life*. Princeton, NJ: Princeton University Press.

Smith, John Maynard. 1965. *The Theory of Evolution*. 2nd ed. Midddlesex: Penguin.

Smithies, Oliver. 2007. Turning pages. PowerPoint presentation for Nobel Lecture. December 7. http://nobelprize.org/nobel_prizes/medicine/laureates/2007 /smithies-lecture.html.

Snow, C. P. 1964. *Two Cultures and a Second Look*. Cambridge: Cambridge University Press.

Sprat, Thomas. 1667. *The History of the Royal-Society of London, for the Improving of Natural Knowledge*. London: J. Martyn and J. Allestry.

Star, Susan L. 1989. *Regions of the Mind: Brain Research and the Quest for Scientific Certainty*. Stanford, CA: Stanford University Press.

Stocking, George, Jr. 1982. *Race, Culture, and Evolution: Essays in the History of Anthropology*. Chicago: University of Chicago Press.

Swales, John M. 1990. Research article in English. In *Genre Analysis: English in Academic and Research Settings*, 137–66. Cambridge: Cambridge University Press.

Swerdlow, N. M. and O. Neugebauer. 1984. *Mathematical Astronomy in Copernicus's "De Revolutionibus,"* part 1. New York: Springer.

Swift, Jonathan. (1735) 1999. *Gulliver's Travels*. New York: New American Library.

Tarsy, D., K. J. Sweadner, and P. C. Song. 2010. Case 17-2010: A 29-year-old woman with flexion of the left hand and foot and difficulty speaking. *New England Journal of Medicine* 362:2213–19.

Taylor, P. J., and A. S. Blum. 1991. Pictorial representation in biology. *Biology & Philosophy* 6 (2): 125–34.

Tilling, Laura. 1975. Early experimental graphs. *British Journal for the History of Science* 8:193–213.

Tufte, Edward R. 1997. *Visual Explanations: Images and Quantities, Evidence and Narrative*. Cheshire, CT: Graphics Press.

———. 2001. *The Visual Display of Quantitative Information*. 2nd ed. Cheshire, CT: Graphics Press.

———. 2003a. *The Cognitive Style of PowerPoint: Pitching Out Corrupts Within*. Cheshire, CT: Graphics Press.

———. 2003b. PowerPoint is Evil, *Wired*, no. 11.09, September. http://www.wired.com /wired/archive//11.09/ppt2.html.

Van Waterschoot van der Gracht, W. A. J. M., ed. 1928. *Theory of Continental Drift: A Symposium on the Origin and Movement of Land Masses Both Inter-Continental and Intra-Continental, as Proposed by Alfred Wegener*. Tulsa: American Association of Petroleum Geologists.

Voss, Julia. 2010. *Darwin's Pictures: Views of Evolutionary Theory, 1837–1874*. Translated by Lori Lantz. New Haven, CT: Yale University Press.

Wainer, Howard. 2005. *Graphic Discovery: A Trout in the Milk and Other Visual Adventures*. Princeton, NJ: Princeton University Press.

Wallace, Alfred Russel. 1905. *My Life: A Record of Events and Opinions*. Vol. 1. New York: Dodd, Mead.

Wang, X., B. Zhang, and D. Z. Wang. 2009. Reductive and transition-metal-free oxidation of secondary alcohols by sodium hydride. *Journal of American Chemical Society* (retracted in 2010).

Watson, James D. 1968. *The Double Helix: A Personal Account of the Discovery of the Structure of DNA*. New York: Athenaeum.

Watson, J. D., and F. H. C. Crick. 1953a. A structure for deoxyribose nucleic acid. *Nature* 171:737–38.

———. 1953b. Genetical implications of the structure of deoxyribonucleic acid. *Nature* 171:964–67.

Wegener, Alfred. 1929. *Die Entstehung der Kontinente und Ozeane*. 4th ed. Braunschweig: Friedrich Vieweg und Sohn.

———. 1966. *The Origin of Continents and Oceans*. Translated by J. Biram. New York: Dover.

Weil, A. A. 1928. Contribution to the pathology of hemichorea. *Brain* 51:36–45.

Weismann, August. 1893. *The Germ-Plasm: A Theory of Heredity*. Translated by W. Parker and H. Ronnfeldt. New York: Charles Scribner's Sons.

West, John B. 2005. Robert Boyle's landmark book of 1660 with the first experiments on rarefied air. *Journal of Applied Physiology* 98:31–39.

Westropp, Hodder M. 1872. *Pre-Historic Phases or, Introductory Essays on Pre-historic Archæology*. London: Bell & Daldy.

Whewell, William. 1984. *Selected Writings on the History of Science*. Edited by Y. Elkana. Chicago: University of Chicago Press.

Whittington, H. B. 1975. The enigmatic animal *Opabinia regalis*, Middle Cambrian, Burgess Shale, British Columbia. *Philosophical Transactions* 271:1–43.

Wilbanks, John. 2006. Another reason for opening access to research. *British Journal of Medicine* 333:1306–1308.

Wimsatt, William C. 2007. *Re-engineering Philosophy for Limited Beings: Piecewise Approximations to Reality*. Cambridge, MA: Harvard University Press.

Wittgenstein, Ludwig. 1953. *Philosophical Investigations*. New York: Macmillan.

Wolin, Richard, ed. 1991. *The Heidegger Controversy: A Reader*. New York: Columbia University Press.

Young, Robert M. 1970. *Mind, Brain, and Adaptation in the Nineteenth Century: Cerebral Localization and Its Biological Context*. Oxford: Clarendon Press.

INDEX

abduction, 129, 130
Alley, Michael, 234–35, 239
Alzheimer's Disease Neuroimaging Initiative (ADNI), 275–77
American Geophysical Society, meeting, 189, 191, 194
analytical philosophy, 11, 18–19, 22, 83
anthropology, visual in, 222–30
antithesis, 219
archaeology, visual in, 222–30
argument: best explanation and, 132–38; construction of, 124–60; in deduction, 127, 162; in dialectical exchange, 267, 269–74, 294; in discovery and justification, 124–26; elegance in, 14, 133, 141; in hypothesis generation, 126–32; in hypothesis pursuit, 132–38; in induction, 127–29, 139, 140, 144, 158, 159, 162, 178–79; model of, 124–27, 159; and narrative, 15, 160–98, 244–50; in PowerPoint, 17, 244–57, 265; in problem selection, 126–27; theory of, 10, 12, 21, 46–47, 128, 250–51
Aristotle, 6
astronomy, visual in, 13, 81, 83–87, 99, 104, 108, 122, 130
Atkinson, Cliff, 237
atlas, 9
Atwater, Tania, 190, 196, 198

Bacon, Francis, 18, 274–75
Baker, David, 283
Barrow, John, 10
Barthes, Roland, 120, 268

Bastide, Françoise, 10–11
Bateson, William, 80
Bazerman, Charles, 7
Bertin, Jean, 10, 272–74, 282, 301, 304
blog, 267, 272–74, 282, 301, 304
Blondlot, Prosper-René, 140
Bolton, J. S., 110–12
bookmarking, 304
Boyle, Robert, 89–91, 286; air pump, 89–91; Boyle's law, 91
Bradley, Jean-Claude, 272–74
Brahe, Tycho, 136
brain localization, 13, 83, 108–20, 122–23
Brownstein, John S., 278, 280–83
bulletin board, 267
Burke, Kenneth, 7

Capecchi, Mario, 231, 233, 235
Carnap, Rudolf, 22
Carpenter, Patricia A., 40–41
Carr, Nicholas, 269
Cassirer, Ernst, 22
Catholic Church, 15
Ceccarelli, Leah, 7
Cesi, Frederico, 13, 81
Chandrasekaran, B., 32–33
chemistry, visual in, 12, 13, 53, 83, 96–102, 108, 122–23
Cicero, 6
citizen volunteers, 274, 282–83
Cleveland, William, 8, 11, 57
cloud chamber, 5, 13, 29, 53, 74, 91
codex, 266–67
collaboration, 137, 267, 274–86, 303, 305